Ecological Impacts of Climate Change

Committee on Ecological Impacts of Climate Change

Board on Life Sciences

Division on Earth and Life Studies

NATIONAL RESEARCH COUNCIL
OF THE NATIONAL ACADEMIES

THE NATIONAL ACADEMIES PRESS

Washington, D.C.

www.nap.edu

THE NATIONAL ACADEMIES PRESS

500 Fifth Street, NW Washington, DC 20001

NOTICE: The project that is the subject of this report was approved by the Governing Board of the National Research Council, whose members are drawn from the councils of the National Academy of Sciences, the National Academy of Engineering, and the Institute of Medicine. The members of the committee responsible for the report were chosen for their special competences and with regard for appropriate balance.

This study was supported by contract/grant no. 08HQGR0005 between the National Academy of Sciences and the U.S. Geological Survey. The content of this publication does not necessarily reflect the views or policies of the U.S. Geological Survey, nor does the mention of trade names, commercial products, or organizations imply endorsement by the U.S. Government.

International Standard Book Number-13: 978-0-309-12710-3
International Standard Book Number-10: 0-309-12710-6

Additional copies of this report are available from the National Academies Press, 500 Fifth Street, NW, Lockbox 285, Washington, D.C. 20055; (800) 624-6242 or (202) 334-3313 (in the Washington metropolitan area); Internet, http://www.nap.edu.

Copyright 2008 by the National Academy of Sciences. All rights reserved.

Printed in the United States of America

THE NATIONAL ACADEMIES
Advisers to the Nation on Science, Engineering, and Medicine

The **National Academy of Sciences** is a private, nonprofit, self-perpetuating society of distinguished scholars engaged in scientific and engineering research, dedicated to the furtherance of science and technology and to their use for the general welfare. Upon the authority of the charter granted to it by the Congress in 1863, the Academy has a mandate that requires it to advise the federal government on scientific and technical matters. Dr. Ralph J. Cicerone is president of the National Academy of Sciences.

The **National Academy of Engineering** was established in 1964, under the charter of the National Academy of Sciences, as a parallel organization of outstanding engineers. It is autonomous in its administration and in the selection of its members, sharing with the National Academy of Sciences the responsibility for advising the federal government. The National Academy of Engineering also sponsors engineering programs aimed at meeting national needs, encourages education and research, and recognizes the superior achievements of engineers. Dr. Charles M. Vest is president of the National Academy of Engineering.

The **Institute of Medicine** was established in 1970 by the National Academy of Sciences to secure the services of eminent members of appropriate professions in the examination of policy matters pertaining to the health of the public. The Institute acts under the responsibility given to the National Academy of Sciences by its congressional charter to be an adviser to the federal government and, upon its own initiative, to identify issues of medical care, research, and education. Dr. Harvey V. Fineberg is president of the Institute of Medicine.

The **National Research Council** was organized by the National Academy of Sciences in 1916 to associate the broad community of science and technology with the Academy's purposes of furthering knowledge and advising the federal government. Functioning in accordance with general policies determined by the Academy, the Council has become the principal operating agency of both the National Academy of Sciences and the National Academy of Engineering in providing services to the government, the public, and the scientific and engineering communities. The Council is administered jointly by both Academies and the Institute of Medicine. Dr. Ralph J. Cicerone and Dr. Charles M. Vest are the chair and vice chair, respectively, of the National Research Council.

www.national-academies.org

COMMITTEE ON ECOLOGICAL IMPACTS OF CLIMATE CHANGE

CHRISTOPHER B. FIELD, *Chair*, Carnegie Institution for Science, Washington, DC
DONALD F. BOESCH, University of Maryland Center for Environmental Science, Cambridge
F. STUART (TERRY) CHAPIN III, University of Alaska, Fairbanks
PETER H. GLEICK, Pacific Institute, Oakland, CA
ANTHONY C. JANETOS, University of Maryland, College Park
JANE LUBCHENCO, Oregon State University, Corvallis
JONATHAN T. OVERPECK, University of Arizona, Tuscon
CAMILLE PARMESAN, University of Texas, Austin
TERRY L. ROOT, Stanford University, CA
STEVEN W. RUNNING, University of Montana, Missoula
STEPHEN H. SCHNEIDER, Stanford University, CA

STAFF

ANN REID, Study Director
FRANCES E. SHARPLES, Director, Board on Life Sciences
ANNE JURKOWSKI, Communications Officer
AMANDA CLINE, Senior Program Assistant

BOARD ON LIFE SCIENCES

KEITH YAMAMOTO, *Chair*, University of California, San Francisco
ANN M. ARVIN, Stanford University School of Medicine, Stanford, CA
RUTH BERKELMAN, Emory University, Atlanta, GA
DEBORAH BLUM, University of Wisconsin, Madison
VICKI CHANDLER, University of Arizona, Tucson
JEFFREY L. DANGL, University of North Carolina, Chapel Hill
PAUL R. EHRLICH, Stanford University, Stanford, CA
MARK D. FITZSIMMONS, John D. and Catherine T. MacArthur Foundation, Chicago, IL
JO HANDELSMAN, University of Wisconsin, Madison
KENNETH H. KELLER, University of Minnesota, Minneapolis
JONATHAN D. MORENO, University of Pennsylvania Health System, Philadelphia
RANDALL MURCH, Virginia Polytechnic Institute and State University, Alexandria
MURIEL E. POSTON, Skidmore College, Saratoga Springs, NY
JAMES REICHMAN, University of California, Santa Barbara
BRUCE W. STILLMAN, Cold Spring Harbor Laboratory, Cold Spring Harbor, NY
MARC T. TESSIER-LAVIGNE, Genentech Inc., South San Francisco, CA
JAMES TIEDJE, Michigan State University, East Lansing
CYNTHIA WOLBERGER, Johns Hopkins University, Baltimore, MD
TERRY L. YATES, University of New Mexico, Albuquerque

STAFF

FRANCES E. SHARPLES, Board Director
JO HUSBANDS, Senior Project Director
ADAM P. FAGEN, Senior Program Officer
ANN H. REID, Senior Program Officer
MARILEE K. SHELTON-DAVENPORT, Senior Program Officer
ANNA FARRAR, Financial Associate
REBECCA WALTER, Senior Program Assistant
AMANDA CLINE, Senior Program Assistant

Preface

The Committee on the Ecological Impacts of Climate Change was given an unusual task; therefore it is appropriate to describe how the committee was formed, how it interpreted its task, and the approach it took to generate this report, so that reviewers and readers are aware of what the report has been designed to achieve. The full statement of task can be found in Appendix A.

The National Research Council (NRC) was approached by the U.S. Geological Survey with a request to produce a scientifically accurate brochure for the general public describing the ecological effects of climate change. Generally, when produced by the NRC, the content of such brochures is derived from previously published NRC consensus reports. In this case, while the NRC has published widely on climate change, the ecological impacts have not been the subject of any recent consensus reports. However, a number of major international consensus reports on climate change, including the Fourth Assessment of the Intergovernmental Panel on Climate Change (IPCC),[1] the Millennium Ecosystem Assessment,[2] several products from the U.S. Climate Change Science Program,[3] and the United Nations Foundation[4] provide ample raw material for such a brochure. Accordingly, the NRC convened a committee of experts to review the published literature and provide a brief report laying out an overview of the ecological impacts of climate change and a series of examples of impacts of different kinds. The contents of

---

[1] IPCC. Climate Change 2007: Synthesis Report. Contribution of Working Groups I, II and III to the Fourth Assessment Report of the Intergovernmental Panel on Climate Change, eds.R. K. Pachauri and A. Reisinger. Geneva: IPCC, 2007.

[2] Millennium Ecosystem Assessment. Ecosystems and Human Well-being: Synthesis. Washington, D.C.: Island Press, 2005.

[3] Backlund, P., A. Janetos, D. Schimel, J. Hatfield, K. Boote, P. Fay, L. Hahn, C. Izaurralde, B. A. Kimball, T. Mader, J. Morgan, D. Ort, W. Polley, A. Thomson, D. Wolfe, M. G. Ryan, S. R. Archer, R. Birdsey, C. Dahm, L. Heath, J. Hicke, D. Hollinger, T. Huxman, G. Okin, R. Oren, J. Randerson, W. Schlesinger, D. Lettenmaier, D. Major, L. Poff, S. Running, L. Hansen, D. Inouye, B. P. Kelly, L. Meyerson, B. Peterson, R. Shaw. The effects of climate change on agriculture, land resources, water resources, and biodiversity in the United States. A Report by the U.S. Climate Change Science Program and the Subcommittee on Global Change Research. Washington D.C.: U.S. Department of Agriculture, 2008.

[4] Scientific Expert Group on Climate Change (Rosina M. Bierbaum, John P. Holdren, Michael C. MacCracken, Richard H. Moss, and Peter H. Raven, eds.). Confronting Climate Change: Avoiding the Unmanageable and Managing the Unavoidable. Report prepared for the United Nations Commission on Sustainable Development. Research Triangle Park, N.C., and Washington, D.C.: Sigma Xi and the United Nations Foundation, April 2007.

this report will be used by the NRC's Division on Earth and Life Studies communications office to design a fully illustrated booklet for distribution to the public.

Members of the committee were chosen to represent knowledge of a wide range of different geographic areas (for example, the arctic or temperate latitudes), and different kinds of organisms and ecosystems. Crucially, in addition to relevant expertise, the committee members were chosen because of their deep familiarity with the international activities that allowed scientists to develop the scientific consensuses on which this report is based and for their experience and skill in conveying complex scientific information to the general public. All eleven committee members served as lead authors on one or more recent scientific assessments on global change and many have been recipients of awards and prizes for exceptional achievement in science communication. The roster of committee members and their biographies are in Appendix B.

The committee met several times by conference call to discuss which examples of the ecological impacts of climate change to provide and how the information should be presented. Because the ultimate audience will be the general public, the committee decided that the report would avoid using jargon and use straightforward examples to help convey complex issues, all while not sacrificing accuracy. At the same time, numerous references and suggestions for further reading are provided for those wishing more detail.

The list of possible examples of ecological impacts of climate change is very long, and only a few can be included in so brief a document. Instead, an effort was made to choose examples from a wide range of ecosystems and of several different kinds of impacts, ranging from range shifts, to seasonal timing mismatches, to indirect consequences of primary impacts. While trying to illustrate the broad range of impacts, the committee also highlighted a few fundamental messages: 1. Climate change and ecosystems are intricately connected and impacts on one will often feed back to affect the other; 2. Ecosystems are complex and their constituent species do not necessarily react to climate change at the same pace or in the same ways; 3. Climate change is not the only stress affecting ecosystems, and other stresses, like habitat loss, overfishing, and pollution, complicate species' and ecosystems' ability to adapt to climate change; 4. These cumulative and interacting changes will likely affect the benefits that humans derive from both managed and unmanaged ecosystems, including the production of food and fiber, purification of water and air, provision of pollinators, opportunities for recreation and much more; and 5. The magnitude of ecological impacts to climate change will depend on many factors, such as how quickly the change occurs; the intensity, frequency, and type of change; and in the long run what actions humans take in response to climate change.

Acknowledgments

This report has been reviewed in draft form by persons chosen for their diverse perspectives and technical expertise in accordance with procedures approved by the National Research Council's Report Review Committee. The purpose of this independent review is to provide candid and critical comments that will assist the institution in making its published report as sound as possible and to ensure that the report meets institutional standards of objectivity, evidence, and responsiveness to the study charge. The review comments and draft manuscript remain confidential to protect the integrity of the deliberative process. We wish to thank the following for their review of this report:

Chad English, SeaWeb, Silver Spring, Maryland
Zhenya Gallon, University Corporation for Atmospheric Research, Boulder, Colorado
Lisa Graumlich, University of Arizona, Tuscon
Richard Hebda, Royal British Columbia Museum, Victoria, Canada
Chris Langdon, University of Miami, Florida
James Morison, University of Washington, Seattle
Robert Twilley, Louisiana State University, Baton Rouge
John Wallace, University of Washington, Seattle
David A. Wedin, University of Nebraska, Lincoln
Donald A. Wilhite, University of Nebraska, Lincoln
Erika Zavaleta, University of California, Santa Cruz

Although the reviewers listed above provided constructive comments and suggestions, they were not asked to endorse the conclusions or recommendations, nor did they see the final draft of the report before its release. The review of this report was overseen by Dr. May Berenbaum of the University of Illinois and Dr. George Hornberger of Vanderbilt University. Appointed by the National Research Council, Drs. Berenbaum and Hornberger were responsible for making certain that an independent examination of this report was carried out in accordance with institutional procedures and that all review comments were carefully considered. Responsibility for the final content of this report rests entirely with the author committee and the institution.

Contents

1	**Introduction**	1
	What are ecosystems and why are they important?	1
	What do we know about current climate change?	3
	What do we expect from future climate change?	11
	Climate change can impact ecosystems in many ways	14
	Ecosystems can adjust to change—over time	15
	Climate Change, other stresses, and the limits of ecosystem resilience	16
2	**Documented Current Ecological Impacts of Climate Change**	17
	Range Shifts	17
	Seasonal Shifts	20
3	**Examples of Ecological Impacts of Climate Change in the United States**	22
	The Pacific Coastline	22
	The Rocky Mountains	25
	The Breadbasket: Central United States	27
	The Northeastern United States	28
	Florida and the Southern United States	29
	The Southwestern Deserts	30
	Alaska and the Arctic	31
4	**Lessons From the Distant Past**	36
5	**Impacts of Future Climate Changes**	38

REFERENCES		41
APPENDIXES		
A	**Statement of Task**	52
B	**Committee Biographies**	53

Introduction

The world's climate is changing, and it will continue to change throughout the 21st century and beyond. Rising temperatures, new precipitation patterns, and other changes are already affecting many aspects of human society and the natural world.

Climate change is transforming ecosystems at extraordinary rates and scales. As each species responds to its changing environment, its interactions with the physical world and the creatures around it change—triggering a cascade of impacts throughout the ecosystem, such as expansion into new areas, the intermingling of formerly non-overlapping species, and even species extinctions

Climate change is happening on a global scale, but the ecological impacts are often local and vary from place to place. To illuminate how climate change has affected specific species and ecosystems, this document presents a series of examples of ecological impacts of climate change that have already been observed across the United States.

Human actions have been a primary cause of the climate changes observed today, but humans are capable of changing our behavior in ways that reduce the rate of future climate change. Human actions are also needed to help wild species adapt to climate changes that cannot be avoided. Our approaches to energy, agriculture, water management, fishing, biological conservation, and many other activities will all affect the ways and extent to which climate change will alter the natural world—and the ecosystems on which we depend.

What are ecosystems and why are they important?

Humans share Earth with a vast diversity of animals, plants, and microorganisms. Virtually every part of the planet—the continents, the oceans, and the atmosphere—teems with life. Even the deepest parts of the ocean and rock formations hundreds of meters below the surface are populated with organisms adapted to cope with the unique challenges each environment presents. In our era organisms almost everywhere are facing a new set of challenges; specifically, the challenges presented by rapid climate change. How have plants, animals, and microorganisms coped with the climate changes that have already occurred, and how might they cope with future changes? To explore these questions we start with a discussion of how plants, animals, and microorganisms fit together in ecosystems and the role of climate in those relationships.

Earth has a great diversity of habitats. These differ in climate, of course, but also in soils, day length, elevation, water sources, chemistry, and many other factors, and consequently, in the kinds of organisms that inhabit them. The animals, plants, and microorganisms that live in one place, along with the water, soils, and landforms, make an ecosystem. When we attempt to understand the impacts of climate change, thinking about ecosystems—and not just individual species—can be helpful because each ecosystem depends on a wide array of interactions among individuals. Some of these involve competition. For example, some plants shade others or several animals compete for the same scarce food. Some involve relationships between animals and their prey. Others involve decomposition, the process of decay that returns minerals and organic matter to the soil. And some interactions are beneficial to both partners, for example, bees that obtain food from flowers while pollinating them.

Climate influences ecosystems and the species that inhabit them in many ways. In general, each type of ecosystem is consistently associated with a particular combination of climate characteristics (Walter 1968). Warm tropical lands with year-round rain typically support tall forests with evergreen broadleaved trees. Midlatitude lands with cold winters and moist summers usually support deciduous forests, while drier areas are covered in grasslands, shrublands, or conifer forests. In a similar fashion shallow tropical-ocean waters harbor coral reefs on rocky bottoms and mangrove forests along muddy shores, whereas temperate shores are characterized by kelp forests on rocky bottoms and seagrasses or salt marshes on sediment-covered bottoms. These major vegetation types or biomes can cover vast areas. Within these areas a wide range of subtly different ecosystems utilize sites with different soils, topography, land-use history, ocean currents, or climate details. Humans are an important part of most ecosystems, and many ecosystems have been heavily modified by humans. A plot of intensively managed farmland, a fish pond, and a grazed grassland are just as much ecosystems as is a pristine tropical forest. All are influenced by climate, all depend on a wide variety of interactions, and all provide essential benefits to people.

The lives of animals, plants, and microorganisms are strongly attuned to changes in climate, such as variation in temperatures; the amount, timing, or form of precipitation; or changes in ocean currents. Some are more sensitive and vulnerable to climate fluctuations than others. If the climate change is modest and slow, the majority of species will most likely adapt successfully. If the climate change is large or rapid, more and more species will face ecological changes to which they may not be able to adapt. But as we will see later, even modest impacts of climate change can cause a range of significant responses, even if the changes are not so harsh that the organism dies. Organisms may react to a shift in temperature or precipitation by altering the timing of an event like migration or leaf emergence, which in turn has effects that ripple out to other parts of the ecosystem. For example, such timing changes may alter the interactions between predator and prey, or plants (including many crops) and the insects that pollinate their flowers. Ultimately we want to understand how climate change alters the overall functioning of the ecosystem and in particular how it alters the ability of the ecosystem to provide valuable services for humans.

Ecosystems play a central role in sustaining humans (Figure 1) (Daily 1997; Millennium Ecosystem Assessment 2005). Ecosystems *provide products* directly consumed by people. This includes food and fiber from agricultural, marine, and forest ecosystems, plus fuel, including wood, grass, and even waste from some agricultural crops, and medicines (from plants, animals and seaweeds). Our supply and quality of fresh water also depends on ecosystems, as they play a critical role in circulating, cleaning, and replenishing water supplies. Ecosystems also *regulate our environment*; for example, forests, floodplains, and streamside vegetation can be critically important in controlling risks from floods; likewise, mangroves, kelp forests, and coral reefs dampen the impact of storms on coastal communities. Ecosystems provide cultural services that *improve our quality of life* in ways that range from the sense of awe many feel when looking up at a towering sequoia tree to educational and recreational opportunities. Ecosystems also *provide nature's support structure*; without ecosystems there would be no soil to support plants, nor all the microorganisms and animals that depend on plants. In the oceans, ecosystems sustain the nutrient cycling that supports marine plankton, which in turn supply food for the fish and other seafood humans eat. Algae in ocean ecosystems produce much of the oxygen that we breathe. In general, we do not pay for the services we get from ecosystems, even though we could not live without them and would have to pay a high price to provide artificially.

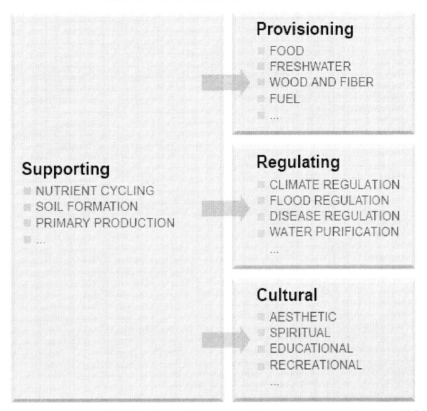

FIGURE 1 Ecosystem services. SOURCE: Millennium Ecosystem Assessment (2005).

Ecosystem services rely on complex interactions among many species, so in most environments it is critical that they contain a diverse array of organisms. Even those services that appear to depend on a single species, like the production of honey, actually depend on the interactions of many species, sometimes many hundreds or thousands. Honey comes from honeybees, but the bees depend on pollen and nectar from the plants they pollinate. These plants depend not only on the bees but also on the worms and other soil animals that aerate the soil, the microorganisms that release nutrients, and the predatory insects that limit populations of plant-eating insects. Scientists are still at the early stages of understanding exactly how diversity contributes to ecosystem resilience—the ability of an ecosystem to withstand stresses like pollution or a hurricane without it resulting in a major shift in the ecosystem's type or the services it provides (Schulze and Mooney 1993; Chapin et al. 1997; Tilman et al. 2006; Worm et al. 2006). But we are already certain about one thing. Each species is a unique solution to a challenge posed by nature and each species' DNA is a unique and complex blueprint. Once a species goes extinct, we can't get it back. Therefore, as we look at the impacts of climate change on ecosystems, it is critical to remember that some kinds of impacts—losses of biological diversity—are irreversible.

What do we know about current climate change?

Over the last 20 years the world's governments have requested a series of authoritative assessments of scientific knowledge about climate change, its impacts, and possible approaches for dealing with climate change. These assessments are conducted by a unique organization, the Intergovernmental Panel on Climate Change (IPCC). Every five to seven years, the IPCC uses volunteer input from thousands of scientists to synthesize available knowledge. The IPCC conclusions undergo intense additional review and evaluation by both the scientific community and the world's governments, resulting in final reports that all countries officially accept (Bolin 2007). The information in the IPCC reports has thus been through multiple reviews and is the most authoritative synthesis of the state of the science on climate change.

Earth's average temperature is increasing
In 2007 the IPCC reported that Earth's average temperature is unequivocally warming (IPCC 2007b). Multiple lines of scientific evidence show that Earth's global average surface temperature has risen some 0.75°C (1.3°F) since 1850 (the starting point for a useful global network of thermometers). Not every part of the planet's surface is warming at the same rate. Some parts are warming more rapidly, particularly over land, and a few parts (in Antarctica, for example) have cooled slightly (Figure 2). But vastly more areas are warming than cooling. In the United States average temperatures have risen overall, with the change in temperature generally much higher in the northwest, especially in Alaska, than in the south (Figure 3). The eight warmest years in the last 100 years, according to NASA's Goddard Institute for Space Studies, have all occurred since 1998 (http://www.giss.nasa.gov/research/news/20080116/).

During the second half of the 20th century, oceans have also become warmer. Warmer ocean waters cause sea ice to melt, trigger bleaching of corals, result in many species shifting their geographic ranges, stress many other species that cannot move elsewhere, contribute to sea-level rise (see below), and hold less oxygen and carbon dioxide.

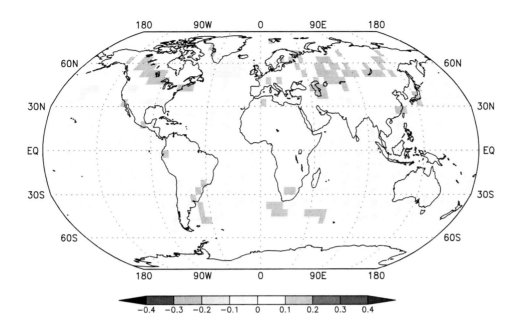

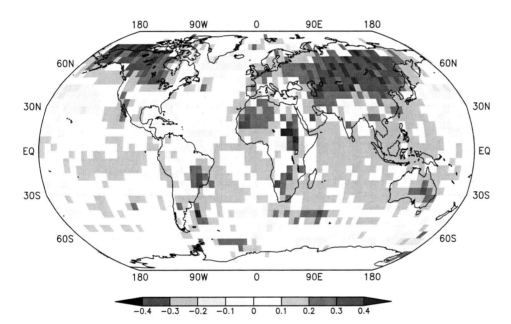

FIGURE 2 Global trends in temperature. The upper map shows the average change in temperature per decade from 1870 to 2005. Areas in orange have seen temperatures rise between 0.1-0.2°C per decade, so that they average 1.35 to 2.7°C warmer in 2005 than in 1870. The lower map shows the average change in temperature per decade from 1950 to 2005. Areas in deep red have seen temperatures rise on average more than 0.4°C per decade, so that they average more than 2°C warmer in 2005 than in 1950. SOURCE: Joint Institute for the Study of the Atmosphere and Ocean, University of Washington.

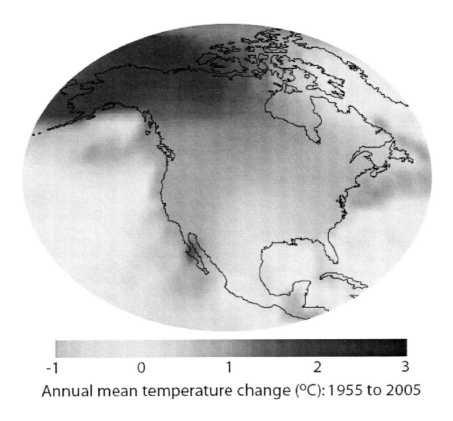

Annual mean temperature change (°C): 1955 to 2005

FIGURE 3 Temperature trends in North America, 1955 to 2005. The darker areas have experienced greater changes in temperature. For example, the Pacific Northwest had average temperatures about 1°C higher in 2005 than in 1955, while Alaska's average temperature had risen by over 2°C. SOURCE: Created with data from Goddard Institute for Space Studies.

Sea levels are rising

Climate change also means that sea levels are rising. Not only do warmer temperatures cause glaciers and land ice to melt (adding more volume to oceans), but seawater also expands in volume as it warms. The global average sea level rose by just under 2 mm/yr (0.08in/yr) during the 20th century, but since satellite measurements began in 1992, the rate has been 3.1 mm/year (0.12in/yr)(IPCC 2007a). Along some parts of the U.S. coast, tide gauge records show that sea level rose even faster (up to 10 mm/yr, 0.39in/yr) because the land is also subsiding. As sea level rises, shoreline retreat has been taking place along most of the nation's sandy or muddy shorelines, and substantial coastal wetlands have been lost due to the combined effects of sea-level rise and direct human activities. In Louisiana alone, 4900 km^2 (1900 mi^2) of wetlands have been lost since 1900 as a result of high rates of relative sea-level rise together with curtailment of the supply of riverborne sediments needed to build wetland soils. The loss of these wetlands has diminished the ability of that region to provide many ecosystem services, including commercial fisheries, recreational hunting and fishing, and habitats for rare, threatened, and migratory species, as well as weakening the region's capacity to absorb storm surges like those caused by Hurricane Katrina (Day et al. 2007). Higher sea levels can also change the salinity and water circulation patterns of coastal estuaries and bays, with varying consequences for the mix of species that can thrive there.

Other effects are being seen

Water Cycle

Climate change is linked to a number of other changes that already can be seen around the world. These include earlier spring snowmelt and peak stream flow, melting mountain glaciers, a dramatic decrease in sea ice during the arctic summer, and increasing frequency of extreme weather events, including the most intense hurricanes (IPCC 2007b). Changes in average annual precipitation have varied from place to place in the United States (Figure 4).

Climate dynamics and the cycling of water between land, rivers and lakes, and clouds and oceans are closely connected. Climate change to date has produced complicated effects on water balances, supply, demand, and quality. When winter precipitation falls as rain instead of snow and as mountain snowpacks melt earlier, less water is "stored" in the form of snow for slow release throughout the summer (Mote 2003), when it is needed by the wildlife in and around streams and rivers and for agriculture and domestic uses. Even if the amount of precipitation does not change, warmer temperatures mean that moisture evaporates more quickly, so that the amount of moisture available to plants declines. The complex interaction between temperature and water demand and availability means that climate change can have many different kinds of effects on ecosystems.

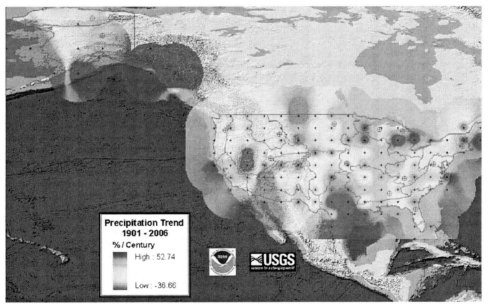

FIGURE 4 Trends in precipitation from 1901 to 2006 in the United States. Areas in red are averaging some 30 percent less precipitation per year now than they received early in the 1900s. Dark blue areas are averaging 50 percent more precipitation per year. SOURCE: Backlund 2008. Created with data from the USGS and NOAA/NCDC.

Extreme Events

The character of extreme weather and climate events is also changing on a global scale. The number of frost days in midlatitude regions is decreasing, while the number of days with extreme warm temperatures is increasing. Many land regions have experienced an increase in days with very heavy rain, but the recent CCSP report on climate extremes concluded that "there are recent regional tendencies toward more severe droughts in the southwestern U.S., parts of Canada and Alaska, and Mexico" (Kunkel et al. 2008, Dai et al. 2004; Seager et al., 2007).

These seemingly contradictory changes are consistent with a climate in which a greater input of heat energy is leading to a more active water cycle. In addition, warmer ocean temperatures are associated with the recent increase in the fraction of hurricanes that grow to the most destructive categories 4 and 5 (Emanuel 2005; Webster et al. 2005).

Arctic Sea Ice

Every year the area covered by sea ice in the Arctic Ocean expands in the winter and contracts in the summer. In the first half of the 20th century the annual minimum sea-ice area in the Arctic was usually in the range of 10 to 11 million km^2 (3.86 to 4.25 million mi^2) (ACIA 2005). In September 2007 sea-ice area hit a single-day minimum of 4.1 million km^2 (1.64 million mi^2), a loss of about half since the 1950s (Serreze et al. 2007). The decrease in area is matched by a dramatic decrease in thickness. From 1975 to 2000 the average thickness of Arctic sea ice decreased by 33 percent, from 3.7 to 2.5 m (12.3 to 8.3 ft) (Rothrock et al. 2008).

Ocean Acidification

About one-third of the carbon dioxide emitted by human activity has already been taken up by the oceans, thus moderating the increase of carbon dioxide concentration in the atmosphere and global warming. But, as the carbon dioxide dissolves in sea water, carbonic acid is formed, which has the effect of acidifying, or lowering the pH, of the ocean (Orr et al. 2005). Although not caused by warming, acidification is a result of the increase of carbon dioxide, the same major greenhouse gas that causes warming. Ocean acidification has many impacts on marine ecosystems. To date, laboratory experiments have shown that although ocean acidification may be beneficial to a few species, it will likely be highly detrimental to a substantial number of species ranging from corals to lobsters and from sea urchins to mollusks (Raven et al. 2005; Doney et al. 2008; Fabry et al. 2008).

Causes of climate change

Both natural variability and human activities are contributing to observed global and regional warming, and both will contribute to future climate trends. It is very likely that most of the observed warming for the last 50 years has been due to the increase in greenhouse gases related to human activities (in IPCC reports, "very likely" specifically means that scientists believe the statement is at least 90 percent likely to be true; "likely" specifically means about two-thirds to 90 percent likely to be true [IPCC 2007b]). While debate over details is an important part of the scientific process, the climate science community is virtually unanimous on this conclusion.

The physical processes that cause climate change are scientifically well documented. The basic physics of the way greenhouse gases warm the climate were well established by Tyndall, Ahrrenius, and others in the 19th century (Bolin 2007). The conclusions that human actions have very likely caused most of the recent warming and will likely cause more in the future are based on the vast preponderance of accumulated scientific evidence from many different kinds of observations (IPCC 2007b). Since the beginning of the Industrial Revolution, human activities that clear land or burn fossil fuels have been injecting rapidly increasing amounts of greenhouse gases such as carbon dioxide (CO_2) and methane (CH_4) into the atmosphere. In 2006 emissions of CO_2 were about 36 billion metric tons (39.6 billion English tons), or about 5.5 metric tons (6.0 English tons) for every human being (Raupach et al. 2007). In the United States average CO_2

emissions in 2006 were approximately 55 kg (120 lb) per person per day. As a consequence of these emissions, atmospheric CO_2 has increased by about 35 percent since 1850. Scientists know that the increases in carbon dioxide in the atmosphere are due to human activities, not natural processes, because they can fingerprint carbon dioxide (for example, by the mix of carbon isotopes it contains, its spatial pattern, and trends in concentration over time) and identify the sources. Concentrations of other greenhouse gases have also increased, some even more than CO_2 in percentage terms (Figure 5). Methane, which is 25 times more effective per molecule at trapping heat than CO_2, has increased by 150 percent. Nitrous oxide (N_2O), which is nearly 300 times more effective per molecule than CO_2 at trapping heat, has increased by over 20 percent (Forster et al. 2007). Scientific knowledge of climate is far from complete. Much remains to be learned about the factors that control the sensitivity of climate to increases in greenhouse gases, rates of change, and the regional outcomes of the global changes. These uncertainties, however, concern the details and not the core mechanisms that give scientists high confidence in their basic conclusions.

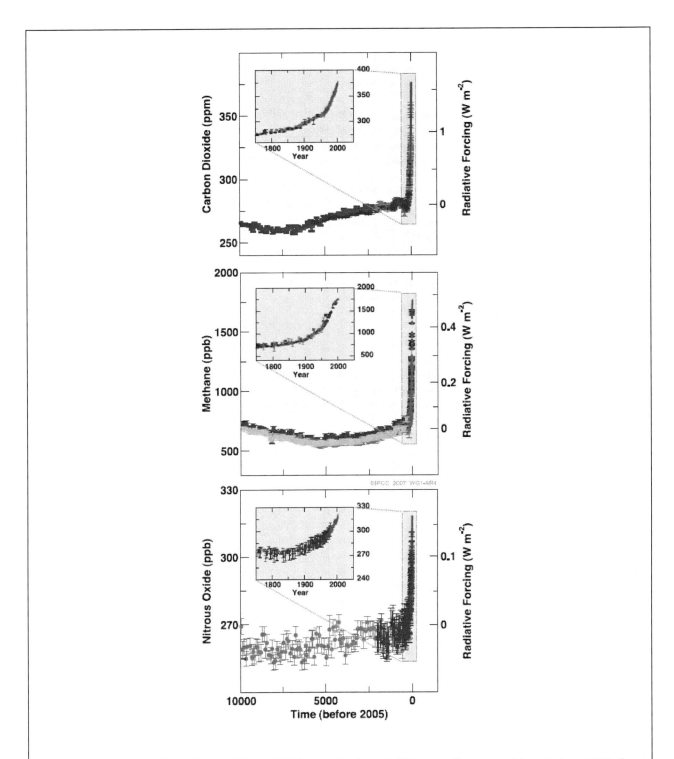

Atmospheric concentrations of CO_2, CH_4 and N_2O over the last 10,000 years (large panels) and since 1750 (inset panels). Measurements are shown from ice cores (symbols with different colors for different studies) and atmospheric samples (red lines). The corresponding radiative forcings (amount of energy trapped per unit area) relative to 1750 are shown on the right hand axes of the large panels. Source: IPCC 2007d.

FIGURE 5: Historical concentrations of greenhouse gasses CO_2, CH_4, and N_2O over the past 10,000 years. For each of these greenhouse gases, the characteristic "hockey stick" shape of the curve is the result of large increases in the concentrations of these gases very recently, compared to their relatively stable levels over the past 10,000 years. SOURCE: IPCC 2007d.

What do we expect from future climate change?

Evidence of rising atmospheric and ocean temperatures, changing precipitation patterns, rising sea levels, and decreasing sea ice is already clear. Average temperatures will almost certainly be warmer in the future. The amount of future climate change depends on human actions. A large number of experiments with climate models indicate that if the world continues to emphasize rapid economic development powered by fossil fuels, it will probably experience dramatic warming during the 21st century. For this kind of "business as usual" future the IPCC (IPCC 2007b) projects a likely range of global warming over 1990 levels of 2.4-6.4°C (4.3-11.5°F) by 2100 (Figure 6, scenario A1F1). If greenhouse gas emissions grow more slowly, peak around the year 2050, and then fall, scientists project a likely warming over 1990 levels of 1.1-2.9°C (2.0-5.2°F) by 2100 (Figure 6, scenario B1).[5]

Temperature increases at the high end of the range of possibilities are very likely to exceed many climate thresholds. Warming of 6°C (10.8°F) or more (the upper end of the projections that the 2007 IPCC rates as "likely") would probably have catastrophic consequences for lifestyles, ecosystems, agriculture, and other livelihoods, especially in the regions and populations with the least resources to invest in adaptation—that is, the strategies and infrastructure for coping with the climate changes. Warming to the high end of the range would also entail a global average rate of temperature change that, for the next century or two, would dramatically exceed the average rates of the last 20,000 years, and possibly much further into the past.

Mean seawater temperatures in some U.S. coastal regions have increased by as much as 1.1°C (2°F) during the last half of the 20th century and, based on IPCC model projections of air temperature, are likely to increase by as much as 2.2-4.4°C (4-8°F) during the present century. "Business as usual" emissions through 2100 would likely lead to oceans with surface temperatures that are 2-4°C (3.6-7.2°F) higher than now and surface waters so acidified that only a few isolated locations would support the growth of corals (Cao et al. 2007). Most marine animals, especially sedentary ones, and plants are expected to be significantly stressed by these changes (Hoegh-Guldberg et al. 2007). Some may be able to cope with either increased temperatures or more acidic waters, but adjusting to both may not be feasible for many species.

---

[5] Projections of warming are given as a range of temperatures for three reasons. First, gaps in the scientific understanding of climate limit the accuracy of projections for any specific concentration of greenhouse gases. Changes in wind and clouds can increase or decrease the warming that occurs in response to an increase in the concentration of greenhouse gases. Loss of ice on the sea or snow on land increases the amount of the incoming sunlight that is absorbed, amplifying the warming from greenhouse gases. Second, the pattern of future emissions and the mix of compounds released to the atmosphere cannot be predicted with high confidence. Some kinds of compounds that produce warming remain in the atmosphere only a few days (Ramanathan et al. 2007). Others, like CO_2, remain for centuries and longer (Matthews and Caldeira 2008). Still other compounds tend to produce aerosols or tiny droplets or particles that reflect sunlight, cooling the climate. Third, there is substantial uncertainty about the future role of the oceans and ecosystems on land. In the past, oceans and land ecosystems have stored, at least temporarily, about half of the carbon emitted to the atmosphere by human actions. If the rate of storage increases, atmospheric CO_2 will rise more slowly. If it decreases, then atmospheric CO_2 will rise more rapidly (Field et al. 2007).

Continued emissions under the "business as usual" scenario could lead by 2100 to 0.6 m (2 ft) or more of sea-level rise. Continuation of recent increases in loss of the ice caps that cover Greenland and West Antarctica could eventually escalate the rate of sea-level rise by a factor of 2 (Overpeck et al. 2006; Meehl et al. 2007; Alley et al. 2005; Gregory and Huybrechts 2006; Rahmstorf 2007).

There will also be hotter extreme temperatures and fewer extreme cold events. An increase in climate variability, projected in some models, will entail more frequent conditions of extreme heat, drought, and heavy precipitation. A warmer world will experience more precipitation at the global scale, but the changes will not be the same everywhere. In general, the projections indicate that dry areas, especially in the latitude band just outside the tropics (for example, the southwestern United States), will tend to get drier on average (IPCC 2007b; Kunkel et al. 2008). Areas that are already wet, especially in the tropics and closer to the poles, will tend to get wetter on average. Increased climate variability and increased evaporation in a warmer world could both increase the risk and likely intensity of future droughts.

Changes in the frequency or intensity of El Niño events forecast by climate models are not consistent (IPCC 2007b). El Niños are important because they are often associated with large-scale drought and floods in the tropics and heavy rains just outside the tropics, but projecting how the interaction between climate change and El Niño events will affect precipitation patterns is difficult. Another example of inconsistent results from models is that model simulations indicate that future hurricane frequency and average intensity could either increase or decrease (Emanuel et al. 2008), but it is likely that rainfall and top wind speeds in general will increase in a world of warmed ocean temperatures.

For all of these different factors—temperature, precipitation patterns, sea-level rise and extreme events—both the magnitude and speed of change are important. For both ecosystems and human activities, a rapid *rate* of climate change presents challenges that are different from, but no less serious than, the challenges from a large *amount* of change (Schneider and Root 2001).

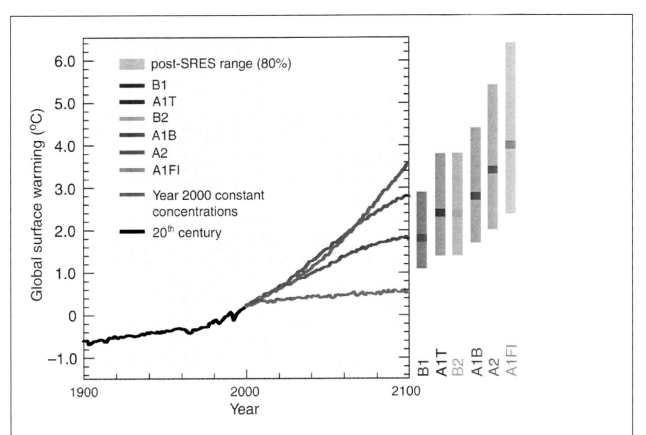

Solid lines are multi-model global averages of surface warming for scenarios A2, A1B and B1, shown as continuations of the 20th-century simulations. These projections also take into account emissions of short-lived GHGs and aerosols. The pink line is not a scenario, but is for Atmosphere-Ocean General Circulation Model (AOGCM) simulations where atmospheric concentrations are held constant at year 2000 values. The bars at the right of the figure indicate the best estimate (solid line within each bar) and the likely range assessed for the six SRES marker scenarios at 2090-2099. All temperatures are relative to the period 1980-1999. SOURCE IPCC 2007b.

FIGURE 6 Projected future temperatures. This figure shows projected trends of average global surface temperature, based on output from all of the major climate models, shown as continuations of the 20th century observations (with the average for 1980-1999 plotted as 0). The pink line represents what would happen if CO_2 concentrations could be held constant at year 2000 levels. Scenarios B1, A1B and A2 represent alternative possible futures. A1B and B1 are futures with modest population growth, rapid economic growth, and a globally integrated economy, with A1B focusing on manufacturing and B1 focusing on service industries. A2 is a world with more rapid population growth but slower economic growth and less economic integration. The bars to the right of the graph represent the likely range of average global temperature from the same models in the years 2090-2099 for a wider range of possible futures, with the horizontal bar in the middle indicating the average across the models. As of 2006, actual CO_2 emissions were higher than those in the A2 scenario, making the full range of scenarios look like underestimates, at least for the first years of the 21st century. (IPCC 2007b, Raupach et al. 2007).

Climate change can impact ecosystems in many ways

Hundreds of studies have documented responses of ecosystems, plants, and animals to the climate changes that have already occurred (Parmesan 2006; Rosenzweig et al. 2007). These studies demonstrate many direct and indirect effects of climate change on ecosystems. Changes in temperature, for example, have been shown to affect ecosystems directly: the date when some plants bloom is occurring earlier in response to warmer temperatures and earlier springs. Extreme temperatures, both hot and cold, can be important causes of mortality, and small changes in extremes can sometimes determine whether a plant or animal survives and reproduces in a given location.

Changes in temperature, especially when combined with changes in precipitation, can have indirect effects as well. For many plants and animals soil moisture is critically important for many life processes; changes in precipitation and in the rate of evaporation interact to determine whether moisture levels remain at a level suitable for various organisms. For fish and other aquatic organisms both water temperature and water flow are important and influenced by the combined effects of altered air temperatures and precipitation. For example, warmer, drier years in the northwestern United States, often associated with El Niño events and anticipated to be more common under many climate scenarios, have historically been associated with below-average snowpack, stream flow, and salmon survival (Mote 2003). Some salmon populations are especially sensitive to summer temperatures; others are sensitive to low stream-flow volumes in the fall (Crozier and Zabel 2006). The fact that climate change leads to rising seas means that organisms and ecosystems located in coastal zones between the ocean and terrestrial habitats are squeezed, especially when the coastal land is occupied by buildings or crops.

The ecological impacts of climate change are not inherently beneficial or detrimental for an ecosystem. The concept that a change is beneficial or detrimental has meaning mainly from the human perspective. For an ecosystem, responses to climate change are simply shifts away from the state prior to human-caused climate change. Measured by particular ecosystem services, some changes could be beneficial; for example, warmer temperatures extend the growing season in some latitudes, and higher CO_2 levels increase the growth of some land plants, with higher potential yields of food and forestry products (Nemani et al. 2003). Others are detrimental, for example, western mountain areas with a longer snow-free season are experiencing increased wildfires, reduced potential wood harvests, and loss of some recreational opportunities (Westerling et al. 2006). In some settings uncertainty about future ecosystem services may be a cost in itself, motivating investments that may not turn out to be necessary or that may be insufficient to effectively address changing needs. To date, many species have responded to the effects of climate change by extending their range boundaries both toward the poles (for example, northward in the U.S.) and up in elevation, and by shifting the timing of spring and autumn events. Plants and animals needing to move but prevented from doing so, for example, because appropriate habitat is not present at higher elevations, are at greater risk of extinction. Shifting species ranges, changes in the timing of biological events, and a greater risk of extinction all affect the ability of ecosystems to provide the critical services—products, regulation of the environment, enhanced human quality of life, and natural infrastructure—they have been providing.

Ecosystems can adjust to change—over time

Ecosystems are not static. They are collections of living organisms that grow and interact and die. Ecosystems encounter an ever changing landscape of weather conditions and various kinds of disturbances, both subtle and severe. Whatever conditions an ecosystem encounters, the individual organisms and species react to the changes in different ways. Ecosystems themselves do not move, individuals and species do; some species can move farther and faster than others, but some may not be able to move at all. For example, a long-lived tree species may take decades to spread to a new range, while an insect with many hatches per year could move quickly. A species that already lives on mountaintops may have nowhere else to retreat. Rapid and extreme disturbances can have major and long-lasting ecological impacts. For example, a severe drought, wildfire, or hurricane can fundamentally reshape an area, often for many decades. In one of the most dramatic examples the impact of an asteroid 65 million years ago is believed to have so radically changed conditions on Earth that the dominant animals, the dinosaurs, died off and were supplanted by mammals (Alvarez et al. 1990).

On longer time scales, most places on Earth have experienced substantial climate changes. During the peak of the last ice age, approximately 21,000 years ago, most of Canada and the northern United States were under thousands of feet of ice (Jansen et al. 2007). Arctic vegetation thrived in Kentucky, and sea levels were about 120 m (400 ft) lower than at present. Over the past million years Earth has experienced a series of ice ages, separated by warmer conditions. Global average temperatures during these ice ages were about 4-7°C (7.2-12.6°F) cooler than present, with the cooling and warming occurring over many thousands of years (Jansen et al. 2007). These ice ages triggered extensive ecological responses, including large shifts in the distributions of plants and animals, as well as extinctions. The massive changes during past ice ages certainly pushed ecosystems off large swaths of Earth's surface as ice-dominated landscapes advanced. However, these changes were generally slow enough that surviving species could move and reassemble into novel, as well as familiar-looking, ecosystems as the ice retreated (Pitelka et al. 1997; Overpeck et al. 2003). The 10,000 years since the last ice age have seen substantial regional and local climate variation, but on a global scale climate was relatively stable, and these regional climate changes did not drive species to extinction nor result in the scale of global ecosystem change seen during glacial-to-interglacial transitions. Even when the global climate is not changing noticeably, regional climate variability (droughts, storms, and heat waves) can have dramatic regional (often short-term) impacts. In a period of climate change it is important to remember that this climate variability will continue to occur on top of the more long-term human-caused climate changes.

Data on ecosystem responses to disturbances in the distant past can provide valuable information about likely responses to current and future climate change. But it is important to recognize that the current rate of increase of CO_2 in Earth's atmosphere is faster than at any time measured in the past, indicating that human-caused global climate change in the current era is likely to be exceedingly rapid, many times faster than the long-term global changes associated with onset and termination of the ice ages (Jansen et al. 2007). One of the big concerns about the future is that climate changes in some places may be too fast for organisms to respond in the ways that have helped sustain ecosystem services in response to natural changes in the past. Understanding how quickly ecosystems can and cannot adjust is one of the key challenges in climate change research.

Climate change, other stresses, and the limits of ecosystem resilience

Climate change is not the only way humans are affecting ecosystems. Humans have a large and pervasive influence on the planet. We use a substantial portion of the land for agriculture and the oceans for fishing (Worm et al. 2006; Ellis and Ramankutty 2008). Many rivers are dammed to provide water for crops or people, or they are polluted with fertilizer or other chemicals. Chemical residues and the by-products of industrial activity, from acid precipitation to ozone, affect plant growth. Human activities, especially land and ocean use, limit some opportunities for species migrations while opening routes for other species. Globally humans have moved many non-native species from one ecosystem to another. Ecosystems operate in a context of multiple human influences and interacting factors.

Earth's ecosystems are generally resilient to some range of changes in climate. A resilient ecosystem is one that can withstand a stress like pollution or rebuild after a major disturbance like a serious storm. A resilient ecosystem can cope with a drought or an unusually hot summer in ways that alter some aspects of ecosystem function but do not lead to a major shift in the type of ecosystem or the services it provides. Thus, a resilient ecosystem may not appear to be affected by modest or slow climate changes. But this resilience has limits. When a change exceeds those limits, or is coupled with other simultaneous changes that cause stress, the ecosystem undergoes a major change, often shifting to a fundamentally different ecosystem type. There is a threshold point when dramatic ecosystem transformations may occur (Gunderson and Pritchard 2002). These thresholds are like the top of a levee as the water level rises. As long as the water level is even slightly below the top of the levee, function is normal. But once it rises above the levee, there is a flood. This kind of threshold response is common in ecosystems, where extreme events like heat waves often serve as triggers for an irreversible transition of the ecosystem to a new state.

Currently plants and animals are responding to rapid climate change while simultaneously coping with other human-created stresses such as habitat loss and fragmentation due to development, pollution, invasive species, and overharvesting. How do we know climate change itself is causing major changes in ecosystems? First, species changing their ranges in the Northern Hemisphere are almost uniformly moving their ranges northward and up in elevation in search of cooler temperatures (Parmesan and Yohe 2003; Parmesan 2006; Rosenzweig et al. 2007). If any or all of the other stressors were the major cause of ecosystem changes, plants and animals would move in many directions in addition to north, and to lower as well as higher elevations. Second, when we look at the association over time of changes between species ranges and temperatures modeled using only natural variation in climate, such as sunspots and volcanic dust in the stratosphere, the relationship is poor. When temperatures are modeled using natural variability as well as human-caused drivers, such as emission of CO_2 and methane, the association is very strong. Consequently, humans are very likely causing changes in regional temperatures to which in turn the plants and animals are responding (Root et al. 2005).

Documented Current Ecological Impacts of Climate Change

Given the compounding factors discussed in the preceding section, it is generally difficult to attribute ecological changes directly or solely to the effects of climate change. Evidence of the ecological impacts of climate change becomes more convincing when trends are observed among hundreds of species rather than relying on studies of a few particular species. Two widely documented and well-studied general ecological impacts of climate change that provide a glimpse into the broader issue are climate-induced shifts in species' ranges and seasonal shifts in biological activities (known as phenology) or events. These types of change have been observed in many species, in many regions, and over long periods of time.

Range and seasonal shifts are not the only general impacts of climate change; other impacts that affect many ecosystems are changes in growth rates, the relative abundance of different species, processes like water and nutrient cycling, and the risk of disturbance from fire, insects, and invasive species.

Range shifts

Climate change is driving the most massive relocation of species to occur without direct human assistance since the beginning of the current interglacial (warm) period (Parmesan 2006). Each species has a range of climates within which it can survive and reproduce. Species can live only in geographic areas where they can tolerate local temperatures, rainfall, and snowfall (see Figure 7). As Earth warms, the tolerable climate ranges for many species are shifting their locations. About 40 percent of wild plants and animals on land that have been followed over decades are relocating in order to remain within suitable climate conditions (Parmesan and Yohe 2003). Maximum range shifts observed during the past 30 years (up to 1000 km poleward and 400 m upward shifts) surpass responses to regional climate variability during the current interglacial (warm) period of the past 10,000 years, and are approaching the magnitudes of range shifts which occurred during the transition from the last glacial maximum to the current interglacial (Coope 1994,1995; Davis and Shaw 2001; Parmesan 2006; Seimon et al. 2007).

Populations or entire species that are unable to move become stressed as the climate around them becomes unsuitable, and ultimately are at high risk of extinction if they cannot relocate (Williams et al. 2003; Thomas et al. 2004; Bomhard et al. 2005; Thuiller et al. 2005; Fischlin et al. 2007). For example, several U.S. Fish and Wildlife Service-listed endangered species live on only one or a few mountaintops. When such a restricted species distribution is coupled with poor dispersal abilities, these species are unlikely to be able to colonize new habitats as their current locations become climatically unsuitable..

One obvious consequence of shifting species ranges is that many of the nature preserves, parks, refuges, and marine protected areas may no longer experience the climates required by the very species for which they were founded. In another hundred years the nation's carefully planned park, preserve, and refuge system may not function as intended (Opdam and Waschler 2004). The movement of species out of the borders of nature preserves is compounded by the fact that some of the preserved areas are also the ones being hardest hit by climate change. For example, the harsh but fragile landscapes of the boreal tundra on the high peaks of the Grand Tetons, the High Sierra, and the Alaska Range, are being strongly affected by human-caused climate change.

Range shifts acutely affect species in the Arctic and Antarctic. Temperatures are rising more rapidly near the poles—up to 3°C (5.4°F) warming since 1850 (compared with 0.75°C [1.3°F] average global increase) (IPCC 2007b). As sea ice gets thinner and shrinks in area, so too shrink animal populations that use ice as their home, including the polar bear and the ringed seal in the Arctic (Stirling et al. 1999; Derocher et al. 2004; Ferguson et al. 2005). In the Antarctic, declines in Adelié penguin populations reflect warming-induced declines in sea ice and warming-induced increases in precipitation (Croxall et al. 2002; Ducklow et al. 2007). These animals are retreating toward the poles, and are rapidly reaching the end of Earth as they know it.

Cold-adapted species living at the tops of mountains are also being stranded with nowhere to move as warmer temperatures—and formerly lower-elevation species—creep up to higher elevations. As these formerly lower-elevation species move into conditions suitable at higher elevations the available land area tends to get smaller as the elevation gets higher (Figure 8). Of course, an upward shift in each forest type means that the next higher type is either eliminated or pushed even higher. The tundra and subalpine plants and animals that grace the tops of the many high peaks and ridges may disappear completely as they are effectively pushed off the tops of the mountains.

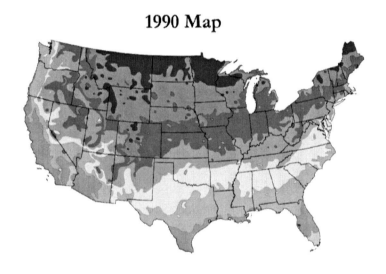

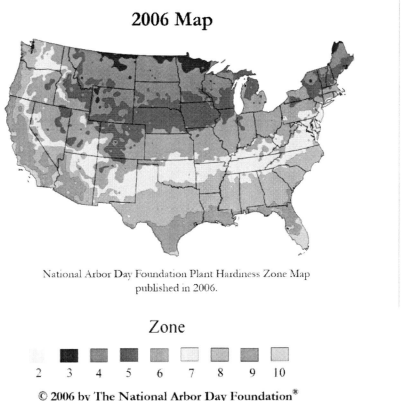

FIGURE 7 Shifts in plant hardiness zones between 1990 and 2006. Many gardeners rely on plant hardiness zones to determine which plants will grow in their region. Each type of plant will thrive only in certain zones. These zones have changed since the map was established. The hardiness zone is moving north in most areas. This means that a plant that once could be grown only in the south can now be grown successfully in areas that were not suitable 15 years ago. However, it also means that some plants can no longer survive where they were planted. SOURCE: The Arbor Day Foundation.

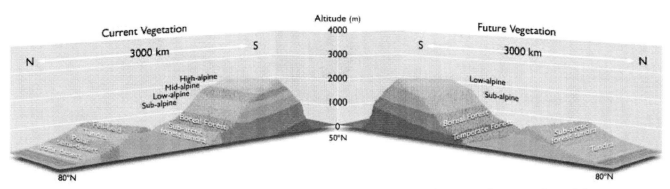

FIGURE 8 This figure shows current and future types of vegetation from north to south and from lower to higher elevation as a result of future warming. Each zone represents a type of ecosystem. In the future these zones move northward but also upward in altitude, replacing existing zones and creating new zones. At an elevation of 1000 m currently one sees subalpine vegetation in the south and fell-field in the north. In a warmer future, at 1000 m one would see

boreal forest in the south and subarctic forest in the north. This process is called range shift. SOURCE: ACIA 2004.

Seasonal Shifts

Climate change is also driving changes in phenology. Many biological events are timed based on seasonal cues, with most of the major ones occurring in the spring and autumn. Many studies looking at changes of the timing of spring events have found that over the last 30 to 40 years, various seasonal behaviors of numerous species now occur 15 to 20 days earlier than several decades ago (Parmesan and Yohe 2003; Root et al. 2003; Parmesan 2007). The types of changes include earlier arrival of migrant birds, earlier appearance of butterflies, and earlier flowering and budding of plants. For example, the date when buds open in the spring in aspen trees in Edmonton, Canada, shifted approximately 26 days earlier between 1900 and 2000, in response to a warming of nearly 2°C (Figure 9) (Beaubien and Freeland 2000). Lilacs carefully observed at over 1100 sites in North America expanded leaves and flowered an average of five to six days earlier in 1993 than in 1959. Autumn changes are not as obvious partly because species vary in the way that earlier springs affect their fall behavior. For example, some birds that arrive earlier in the spring also leave earlier in the fall, regardless of the weather. Many trees, on the other hand, respond to a later arrival of fall by delaying the date their leaves turn color.

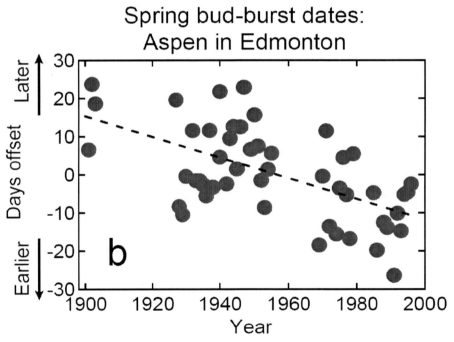

FIGURE 9 This graph shows when the buds on aspen trees opened in Edmonton, Canada during the 20th century. The zero point is the average date (for the entire century) when buds opened. Each circle represents an historical record of when buds opened in that particular year. The dotted line shows the trend; aspen buds are opening on average 25 days earlier than they did a century ago in response to warmer temperatures. The change in blooming date is an example of a seasonal, or phenology, shift. SOURCE: adapted from data in Beaubien and Freedland (2000).

If all the different species in an ecosystem shifted their spring behavior in exactly the same way, the impact of warming temperatures might be minimal. But what happens when a species depends upon another for survival (predator on prey, for example) and only one changes the timing of its spring activity? Such a change can disrupt the predator-prey interaction, which in turn can cause a drop in the predator population. For example, in Europe the bird known as the pied flycatcher has not changed the time it arrives on its breeding grounds, but the caterpillars it feeds its young are emerging earlier (Both et al. 2006). Missing the peak of food availability means fewer chicks are surviving and the pied flycatcher population is declining.

Another example of mismatched predator-prey emergence is seen in plankton blooms in the North Sea near England. There, many kinds of plankton (small marine organisms) have changed the timing of their major blooms, but not by the same amount. In response to a warming of about 0.9°C (1.6°F), *Ceratium fusus*, a tiny plant-like organism, shifted its peak bloom about a month earlier in 1981-2002, compared to 1958-1980, but copepods, their shrimp-like predators, shifted by only 10 days. This kind of mismatch appears to be common in the North Sea, with plants generally shifting farther than the animals that feed on them (Edwards and Richardson 2004).

Examples of Ecological Impacts of Climate Change in the United States

Climate change is global in scope, but ecological impacts are often quite localized. Although most of the evidence of the ecological impacts of climate change stems from trends observed among hundreds of species rather than a particular species, there are compelling examples of how climate change has affected individual species and ecosystems. The following examples review just a few of the ecological changes that have been documented in regions across the United States. Future projections of the effects of climate change on these areas are also explored, although it should be noted that such projections are based on the continuation of current trends in anthropogenic contributors to climate change. If human activities change, so too may these projections.

The Pacific Coastline

Edith's and Quino checkerspot butterfly
We know some species are very sensitive to climate which allows them to act as early warning indicators for climate change. One such species is Edith's checkerspot butterfly (*Euphydryas editha*), a species with a marked range shift over the past 100 years that has been attributed to climate change.

Forty years of research have documented strong responses of wild populations of Edith's checkerspot butterfly to the vagaries of weather and to climates with strong seasonal variation. Weather extremes cause local extinctions but this is a natural part of Edith's checkerspot biology (Singer & Ehrlich 1979, Singer & Thomas 1996). Using museum records to determine where Edith's checkerspot lived in the past, an asymmetrical pattern of population extinctions on a continental scale was revealed. Population extinctions were four times as high at the southern end of the butterflies' range (in Baja, Mexico) than at the northern end (in Canada), and nearly three times as high at lower elevations (below 2400 m (8,000 ft)) than at higher elevations (from 2400 to 3800 m (8,000 to 12,500 ft)) (Parmesan 1996). This extinction process has effectively shifted the range of *E. editha* both northward and upward in elevation since the beginning of the 20th century—a shift in concert with temperature increases resulting from climate change.

Separate analyses showed that other factors (such as proximity to large urban areas) were not associated with the observed extinction patterns. Since the only strong associations were between the extinction patterns and various climate trends, regional climate warming was by default the most likely cause of the observed shift in the butterfly's range.

The Quino checkerspot (*E. editha quino*) is a federally listed endangered subspecies of Edith's checkerspot whose case highlights the conservation implications of climate change. Although habitat destruction is the primary cause of the decline of the Quino checkerspot, climate change poses problems for its recovery. Quino checkerspot populations along the southernmost range (in Mexico) face the lowest degree of threat from development. Unfortunately, these habitats are at the greatest risk from continuing warming and drying climate trends. By contrast, Quino habitat that might have been available farther north has been destroyed by development in the Los Angeles/San Diego corridor. The case of the Quino checkerspot has resulted in the first habitat recovery plan to list climate change not only as a current threat but also as a factor that should be considered in designing habitat reserves and recovery management (Anderson et al. 2001).

Pacific Ocean and fisheries
With seafood providing the primary source of protein for more than 1 billion people worldwide, and demand for seafood growing exponentially, the future of the world's fisheries is of critical importance. There is, however, very limited understanding of how global climate change might affect whole ocean ecosystems. Some of what we have learned about how changes in climate affect marine ecosystems comes from what has been observed during periodic climate cycles, such as the El Niño-Southern Oscillation, the Pacific Decadal Oscillation, and the North Pacific Gyre Oscillation, natural climatic fluctuations generated by ocean-atmosphere interactions over the Pacific that can have important effects on weather conditions globally For example, some species' distributions change with El Niño cycles. Other ecosystem changes, however, appear to be unlinked to those cycles, and instead seem to have arisen as novel, unexpected perturbations.

One such anomaly is a new dead zone that has appeared off the Pacific Northwest coasts of Washington and Oregon. A dead zone is an area of the ocean with insufficient oxygen to support most marine life. Most animals that cannot swim or scuttle away suffocate. The zone of low (or no) oxygen along the Pacific Northwest is different from most of Earth's other approximately 400 dead zones that are caused by runoff of excess nutrients from the land, usually from agricultural lands (Diaz and Rosenberg 2008). The Pacific Northwest dead zone first appeared in the summer of 2002 and has appeared each summer since that time.

This dead zone is not a result of fertilizer use or sewage from land-based sources. Its ultimate cause is still under investigation, but three immediate causes have been documented, each of which is possibly linked to climate change. (1) Warmer ocean waters hold less oxygen at the surface and slow the resupply of oxygen to deeper waters (Stramma et al. 2008). (2) Changes in the coastal winds that control a process called coastal upwelling have been documented (Barth et al. 2007). (3) Changes in ocean circulation that bring waters with abnormally low oxygen and high nutrient levels to the surface during upwelling have been recorded. Reliable and comparable oxygen measurements in this coastal ocean date back 60 years. Researchers who compared dissolved oxygen content in coastal waters along the Oregon shore conclude that the recent seven years (starting in 2002) have been dramatically different from the previous fifty years (Chan et al. 2008). Analysis of over 10,000 individual dissolved oxygen measurements indicate that prior to the early 2000s only one record showed severe hypoxia (low oxygen of less than 0.5 ml of oxygen per liter of water), and none showed anoxia (no oxygen) in nearshore coastal waters.

Starting in 2002 the dead zone has appeared repeatedly each summer. The most severe low-oxygen event on record was in 2006 along the Pacific Northwest coast; that dead zone lasted four months and occupied up to two-thirds of the water column, oxygen levels dropped to zero, and there were widespread die-offs of seafloor life that could not get away quickly enough. The implications for fisheries in the region are under investigation.

Other ecosystems off the coasts of Peru, Chile, Namibia, South Africa, and Morocco also seem to be undergoing changes involving dead zones, although the specifics vary across these systems. These large marine ecosystems all depend on coastal upwelling; they collectively represent 1 percent of the surface area of oceans, but have historically provided around 20 percent of the fisheries (Pauly and Christensen 1995). If catches diminish and ecosystem functioning is disrupted as a consequence of emerging or expanding dead zones, the consequences for many of the world's key fisheries could be substantial.

Wine Quality in California

Climate change affects managed ecosystems like vineyards just as it affects natural ecosystems, and thereby can have major economic and social effects. Wine is one of California's most important agricultural products. The industry takes in billions of dollars per year and is a critical part of the State's cultural fabric. Wine grapes can grow in a wide range of climates and soils, but the quality of the crop and its value for producing high-quality wine depends on something the growers call "terroir," a subtle balance of climate, soils, and landforms. Terroir can be so important that the price of grapes from a premium wine region typically fetch more than 10 times the price of the same variety grown elsewhere. Climate changes from 1950 to 1997 generally improved conditions for growing grapes in California's premium wine regions (Nemani et al. 2001). A modest warming, especially at night, decreased the incidence of frost and advanced the start of the growing season. Further warming would, however, be unlikely to aid the industry. One study concluded that the warming associated with "business as usual" emissions would, by the last decades of the 21st century, degrade California's premium wine regions to marginal from their current status of optimal (Figure 10) (Hayhoe et al. 2004). Another study concluded that the area with the potential to produce premium wines could decrease by up to 81 percent (White et al. 2006).

		2020-2049				2070-2099			
	1961-1990	PCM model		HadCM3 model		PCM model		HadCM3 model	
Wine Grape growing region		lower emissions B1	higher emissions A1fi	lower r emissions B1	higher emissions A1fi	lower emissions B1	higher emissions A1fi	lower emissions B1	higher emissions A1fi
Wine Country	Optimal (mid)	Optimal (mid)	Optimal (high)	Optimal (mid)	Optimal (high)	Impaired	Impaired	Marginal	Impaired
Cool Coastal	Optimal (low)	Optimal (mid)	Optimal (mid)	Optimal (mid)	Optimal (mid)	Optimal (mid-high)	Optimal (high)	Optimal (mid-high)	Impaired
Central Coast	Optimal (mid-high)	Optimal (high)	Optimal (high)	Optimal (high)	Optimal (high)	Marginal	Marginal	Marginal	Impaired
Northern Central Valley	Marginal	Impaired	Impaired	Impaired	Impaired	Impaired	Impaired	Impaired	Impaired
Southern Central Valley	Impaired	Impaired	Impaired	Impaired	Impaired	Impaired	Impaired	Impaired	Impaired

FIGURE 10 Projected conditions for grape growing in certain regions of California, from two different climate models, the low sensitivity PCM model and the medium sensitivity HadCM3 model. Each model was used to project conditions for grape growing under conditions of higher or lower CO_2 emissions. Both models project that conditions will improve in some regions in the medium term (2020-2049), but that by 2070-2099 only the cool coastal regions will still have optimal conditions for growing grapes (Hayhoe et al. 2004).

Marine species along the Pacific coastline
Changing climate has already affected the distribution of marine organisms from south to north along the coasts. For example, in one long-term study of animals and plants that inhabit rocky shores along the central California coasts, many southern species became more common while northern species became more rare over the 60-year period ending in the mid-1990s (Sagarin et al. 1999). During that period, shoreline ocean temperatures increased by 0.8°C (1.4°F) and summer maximum temperatures by nearly 2.2°C (4°F). Other changes in species abundances that are consistent with expectations of climate change have been reported in this system (Smith et al. 2006) as well as for rocky shores in Europe (Southward et al. 2005). Very rapid shifts in geographic distribution have been recorded for bottom-dwelling species that are important to fisheries in the Bering Sea (Mueter and Litzow 2008). A clear northerly migration of snow crab, rock sole, halibut, and pollock has been reported with rates of movement 2-3 times faster than the average rate found for terrestrial species (Parmesan and Yohe 2003). These species appear to be shifting northward in response to the northward movement of the extent of seasonal ice.

The Rocky Mountains

Range of the American pika
Once havens for cold-adapted species, mountaintops around the world are now showing signs of warming stress. Consider the case of the American pika, a small mammal that looks like a hamster but is actually more closely related to rabbits. Paleoecological records show that it lived in the lowlands during the last ice age. As the ice retreated, the pika that had once lived across the entire landscape gradually shifted uphill—an easy move. It now survives in isolated mountaintop islands on various mountain ranges throughout the western United States. Populations below about 7000 ft are rapidly going extinct, with past physiological studies suggesting stress from high temperature is the cause (Beever et al. 2003; Smith 1974).

Trout habitat
Earlier springs and warmer summers are beginning to restrict trout habitat severely in some of the small headwater streams of the Rocky Mountains, home to legendary trout fisheries. As snowpacks melt earlier in the spring, late summer streamflows of cool snowmelt water are declining, and some small rivers, like the Big Hole in Montana, cease to flow in late summer, becoming isolated pools until replenished by fall rains. Trout die at water temperatures above 26°C (78°F), and some stream temperatures are now reaching lethal levels in July and August. If current trends continue, coldwater species like trout could increasingly be restricted to the most permanent streams. Late summer stream flow in seven Montana rivers has dropped an average of 30 percent since 1950 as a result of increasing irrigation demand, earlier snowmelt, and warmer summer temperatures. State officials have had to temporarily close recreational trout fishing during August in recent years on certain streams because of low stream flow and high water temperatures. From 18 to 92 percent of bull trout habitat could be lost in the northern Rocky Mountains in the next half century due to global warming influences on stream temperatures (Rieman et al. 2007). Rocky Mountain lakes will likely see an increase in the abundance of warmwater fish like yellow perch and smallmouth bass but a decrease in coldwater species.

Spring emergence of yellow-bellied marmots
High in the Colorado Rockies, some important members of the animal community are responding to warmer springtime temperatures. The yellow-bellied marmot, a large burrowing mammal of the squirrel family, very common in the western mountains of the United States, emerged from hibernation 23 days earlier (around April 1) in 1999 than in 1976, apparently in response to warmer late-winter temperatures. This unfortunately means they are present and active when snow still covers their normal food; in this Rocky Mountain location warmer temperatures were not, by 1999, associated with earlier melting of the winter snowpack. Cut off from food the marmots need to survive longer on stored reserves, potentially decreasing their ability to reproduce (Inouye et al. 2000).

Forests: a deadly combination of drought, wildfire, and insect pests
In much of the country winter temperatures are not as severe as before, which may be more convenient for humans but throughout the western mountains we now see more wintertime precipitation falling as rain instead of snow (Knowles 2006). Thus the winter snowpack, which is crucial for summer water resources, is no longer providing as much free natural storage as before. Winter snowpack used to peak around April 1, when snow hydrologists would take a measure of peak snow depth for each year's water management planning. Now the April 1 snowpack is 10 to 20 percent lower than 50 years ago, partly because of less snow but also because the spring melt begins on average 2 to 4 weeks earlier (Mote et al. 2005; Barnett et al. 2008).

In climates with adequate summer rainfall, earlier springs mean a longer growing season, which may result in greater plant growth. But in the arid climates of the West, earlier snowmelt and warmer spring temperatures mean that the annual summer drought may now begin in the spring. Western valleys have always been dry, but now the higher mountain forests, from 1200 m (4000 ft) and higher in Montana to over 3600 m (12,000 ft) in Arizona are no longer as protected by a slowly melting snowpack as before. The longer, more intense spring-summer drought is stressing mountain forests (Logan et al. 2003).

Wildfire occurrence and extent are also dramatically escalating in western forests (not only in the U.S. Rockies but also in western Canada and Alaska), a legacy of both a changing climate and decades of total fire suppression that has resulted in a dramatic buildup of dead fuels. In the last 20 years the western fire season has expanded by over 2.5 months (Westerling et al. 2006). In 2007 California had wildfires burning in November, and Billings, Montana, had a wildfire burning on January 8, 2008. There are now four times as many wildfires exceeding 4 km^2 (1.5 mi^2) as there were 30 years ago, and these frequent large fires are burning six times as much area. The national wildfire-fighting budget (already strained for other reasons such as increased development in fire vulnerable areas) now exceeds $1 billion almost every year, and an ominous interaction is now emerging: as insect epidemics kill vast areas of forest, they leave standing dead fuels for even larger wildfires. Ecologists now expect that some of these areas will not recover as forests; they will instead return as more open savannah or grassland ecosystems. The climate is becoming too dry to support some of our nation's forests.

An unprecedented bark beetle epidemic has affected 47,000 km^2 (18,000 mi^2) of forest in western North America over the last 10 years (Raffa et al. 2008). The epidemic is an example of the complicated interactions that characterize ecosystem dynamics: in this case the interplay between a changing climate, vulnerable trees, and opportunistic insects. Mild overwintering temperatures have allowed more insect larvae to survive the winter (Logan et al. 2003). At the same time longer, dryer, and warmer, summers have both accelerated beetle life cycles and stressed the trees upon which these beetles feed. The stressed trees produce less pitch, which makes them more susceptible to beetle damage. Many of the affected forests are made up of trees that are all the same age because of earlier wildfire damage. These uniform forest age structures provide contiguous landscapes allowing rapid population dispersal and successful beetle attacks on the stressed trees.

The Breadbasket: Central United States

Agricultural Shifts
The central part of the United States is one of the world's great agricultural regions. With rich soils and a favorable climate, the region produces some of the world's highest yields of corn, soybeans, and wheat. Production of corn and soy beans are centered east of the plains, with the highest production in Iowa, Illinois, Minnesota, and Indiana. Wheat is grown mainly in the western part of the region, especially in the Dakotas and Kansas. For these three crops, yields have increased steadily over the last century, reflecting the combined effects of improved seed stocks and improved management (National Agricultural Statistics Service 2008).

Continuing efforts by farmers and scientists to increase yields make it difficult to assess whether climate changes to date have had an effect. In general, the rate of plant growth increases with warming, up to a point, but it decreases when temperatures get above that point. For each crop, varieties have adapted or been bred to thrive in a range of temperatures, but there are limits to the temperature range for each crop. When temperatures get too warm, crops tend to mature early, completing growth before the end of the season. Under extreme conditions crops can be killed by either high or low temperatures. Projecting crop growth in a changing climate is further complicated by the fact that for most plant species, growth increases in response to an increase in the CO_2 concentration of the atmosphere. Plants grow by combining CO_2 from the atmosphere with water, using energy from light, to make carbohydrates, in the process called photosynthesis. Under open-field conditions, increasing atmospheric CO_2 concentrations (but not changing temperature) from the current ambient level to a level that may occur in 30 to 50 years increases the yield of plants like soybeans or wheat an average of about 15 percent (Long et al. 2006). Plants like corn and sugarcane have a mechanism that concentrates CO_2 in the leaves, and they generally do not grow more when exposed to elevated CO_2. Unhappily, some pest plants (like poison ivy) grow faster and produce more of their characteristic irritant when atmospheric CO_2 concentration is higher (Mohan 2006).

A number of models estimate changes in crop yields in response to changing climate and CO_2 concentrations. At the global scale these models suggest that if warming is modest, yields in hot regions will decrease, while yields in cool regions will increase. For the temperate Great Plains, most models conclude that warming up to $2°C$ will probably increase average yields by 5 to 20 percent (Field et al. 2007b). Depending on the amount of warming that occurs, these yield increases may persist through the century.

The actual impact of climate change on crop yields in the future depends on a number of factors. The balance between effects of warming (which can increase or decrease yields) and effects of increased CO_2 (which increases yields for some crops but not for corn) will mean increased yields for some crops and different varieties within crops and decreases for others. Higher temperatures can interfere with pollination and seed set, resulting in reductions in productivity; the acceleration of plant life cycles can result in crops that are smaller when they mature (Hatfield et al. 2008). Effects of climate change on the competitive ability of weeds and other pests, on the susceptibility of weeds to herbicides, and on the frequency of severe weather all potentially complicate the challenge of sustaining historical rates of yield increases (Hatfield et al. 2008). Another critical factor in future yields is the level of adaptation by farmers. Aggressive action to adjust farming methods, planting dates, and the crops or varieties grown can play a large role in yields. The potentially large contrast between yields with and without farmers' taking steps towards adaptation underscores the importance of good information for coping effectively with climate change.

Migratory Waterways
The natural ecosystems in the central United States are also affected by climate change. Millions of migratory birds fly back and forth across the central United States, many of them resting, feeding, and mating in temporary lakes called playa lakes in the south and prairie potholes in the north. The region is especially critical for mallard ducks and other waterfowl, with their annual population numbers corresponding closely to the number of these temporary wetlands available at the beginning of the breeding season. The health of the prairie potholes for waterfowl habitat depends on whether future precipitation increases sufficiently to offset warmer temperatures. A combination of higher temperatures and lower rainfall could dry up potholes in a region covering six U.S. states and three Canadian provinces, home to the most productive waterfowl habitat on Earth (Johnson et al. 2005). These shallow water holes are already under pressure because they are used to provide water for irrigation, filled to provide more land for crops and houses, and often subject to runoff of nutrients and pesticides. Climate change could further stress these essential but transient habitats (Covich 1997).

The Northeastern United States

Northeast Fisheries
New England fisheries have for many decades been based on cod and lobster. Stocks of Georges Bank cod, flounder, and haddock have collapsed due to overfishing. It is increasingly clear that the cod fishery is also vulnerable to stress related to warming (Fogarty et al. 2008). Cod require average bottom-water temperatures cooler than 12°C (54°F) and cooler than 8°C (46°F) for growth and survival of young. If future warming of bottom waters is limited to the low end of projected increases by the end of this century, cod may still survive over much of its current range from Long Island to the Gulf of Maine, but more substantial warming will likely push temperatures south of Cape Cod above the 12°C (54°F) threshold. Cod could survive in cooler pockets north of Cape Cod and the cooler, historically rich waters of Georges Bank.

Lobsters tolerate a wider range of water temperatures, however in warmer water lobsters need more oxygen to survive, and warmer water holds less oxygen. As temperatures approach 26°C (79°F) the concentration of oxygen in the water becomes insufficient for lobsters. Since the late 1990s, lobster populations in Long Island Sound have fallen rapidly, with harvests in 2003 only 20 to 30 percent of their earlier size. While many factors may contribute to this decline, warming is probably part of the mix, as water temperatures have exceeded 26°C with increasing frequency (Frumhoff et al. 2007). In the Gulf of Maine warmer conditions in the future will probably improve lobster habitat, providing a longer growing season, more rapid growth, and more area suitable for the growth and survival of juveniles. However, such warmer temperatures may have other indirect effects on lobsters. Since the late 1990s, lobster-shell disease, caused by a bacterium, has been prevalent in southern New England, making the affected lobsters unmarketable. While temperature increases have not been demonstrated to be the cause of greater disease prevalence, the spatial pattern of disease occurrence suggests temperature is a factor (Glenn and Pugh 2006).

We are also seeing animal parasites moving northward. For example, the oyster parasite *(Perkinsus marinus)* extended its range northward from the Chesapeake Bay to Maine, a 500 km (310 mi) shift with potentially major implications for oyster fisheries. Censuses from 1949 to 1990 showed a stable distribution of the parasite from the Gulf of Mexico to its northern boundary at the Chesapeake Bay. A rapid northward expansion of the parasite in 1991 has been linked to above-average winter temperatures, rather than to human-driven introduction or genetic change (Ford 1996).

Florida and the Southern United States

Northward movement of tropical species
Just as cold-adapted species are moving toward the poles and up the mountains, many tropical species are also moving into the United States. Former migrants like the rufous hummingbird (*Selasphorus rufus*) and the Mexican green jay *(Cyanocorax yncas)* have become year-round residents in Alabama and Texas, respectively (Hill et al. 1998). Florida has five new species of tropical dragonfly (Paulson 2001). Many tropical butterflies that are normally confined to Mexico are starting to breed as far north as Austin, Texas.

Having these new species in U.S. backyards has delighted bird and butterfly watchers, and overall species diversity may have actually increased along the southern U.S. border. The observed northward movement of species also has such potentially negative implications as new arrivals competing with local species for scarce resources, bringing with them new diseases, or crowding out native species.

Sea-level rise and the Everglades
The Florida Everglades are a unique ecosystem: a vast subtropical wetland of sawgrass, mangrove forests, and cypress swamps that are home to wading birds, alligators, wood storks, Florida panthers, and manatees. Beginning in the late 1800s, first to support agriculture and later to protect a rapidly growing human population from flooding, a large part of the natural wetlands have been drained or otherwise managed. These activities shrank the Everglades to half their original size, fresh water flows through the wetlands changed dramatically, pollution from urban areas and fertilizer runoff from agricultural areas increased, and saltwater from the surrounding oceans infiltrated further inland.
In response to the deterioration of this natural resource—the Everglades National Park is a designated International Biosphere Reserve, a World Heritage Site and a Wetlands of International Importance—a number of preservation and restoration initiatives are underway. Sea-level rise has already resulted in the landward encroachment of salt-tolerant mangroves (Ross et al. 2000) and the task of restoring the Everglades will be made more difficult by the higher rate of sea-level rise, increase in water temperature, changes in precipitation, and more extreme storms that are expected to result from climate change (Twilley et al. 2001; NRC 2008).

In regions of the low-lying southeastern U.S. coast that are undergoing subsidence, a projected relative sea-level rise of 0.6 to 1.2 m (2 to 4 ft) over the 21st century would reconfigure shorelines and fragment barrier islands. Coastal saltmarshes and mangroves would be hard pressed to accumulate new soil fast enough to keep pace with the rising water level, and many of them would be lost. The migration of wetlands inland as sea levels rise is an important means of coastal ecosystem adaptation. When human development prevents such migration, adaptation is more difficult. Conversely, protection of regions into which coastal wetlands can migrate would ease adaptation. Even where landward migration is not blocked by development, rapid climate change would make it unlikely that this process could happen fast enough to compensate for the losses. Animal species that depend on coastal marshes and mangroves, ranging from fish species that use them as nursery areas to migratory waterfowl and wading birds will likely be adversely affected.

Coral reefs
Coral reefs provide many ecosystem services including acting as a habitat for many kinds of fish and a protective barrier for nearby shores. Reefs off the Florida Keys and in other tropical U.S. waters are to varying degrees already in degraded condition due to the effects of overfishing, land-based pollution by nutrients and sediments, and coastal development (Pandolfi et al. 2005). Ocean warming and acidification due to increasing carbon dioxide concentrations pose an additional double threat that will challenge the survival of coral reefs (Hoegh-Guldberg et al. 2007). Heat stress in shallow tropical waters causes corals to expel the colorful symbiotic algae that provide a primary source of nutrition for the coral, leaving only the white "bone structure" of the corals behind. This process, called coral bleaching, can be lethal to the coral if it lasts too long. Bleaching has increased in intensity and frequency in recent decades. Bleaching and coral mortality become progressively worse as unusually high temperatures are experienced over longer periods. Increased acidity is likely to slow or stop the growth of coral over this century, making them less competitive with the seaweeds that overgrow them, and reducing the capacity of corals to build reefs. Not only are the corals themselves in jeopardy but so is the survival of the myriad species found only on coral reefs, which make coral reefs one of the most diverse ecosystems on Earth.

The Southwestern Deserts

Wildfire and invasive species
Until recently the Mojave and Sonoran deserts of the southwestern United States were generally fireproof. There simply was not enough fuel to carry a fire from shrub to shrub or cactus. However, a number of non-native grasses have now become successfully established in much of the area, transforming fireproof desert into highly flammable grassland. Important examples include buffelgrass (*Pennisetum ciliare*) a non-native grass originally from Africa that is spreading rapidly over large parts of the Sonoran desert, and other grasses (e.g., *Bromus rubens*) in the Mojave (Brooks and Matchett 2006). Like many fire-*prone* grasses, these are also fire-*adapted*, sprouting again quickly and densely following fire, pushing out the native species, including the iconic Saguaro cactus, which is not adapted to frequent fire (Esque et al. 2004). While climate change is not implicated in the spread of buffelgrass, there is concern that warming temperatures will allow this plant to continue to thrive in the desert Southwest, and also extend its range to higher elevations.

The piñon pine
We tend to think of established vegetation, especially in desert regions, as very drought tolerant, but drought tolerance has limits. Severe drought, especially combined with warming, has the potential to push ecosystems past those limits. That is exactly what happened recently across much of the Four Corners region, where New Mexico, Arizona, Colorado, and Utah meet. This region experienced a severe drought from 2000 to 2003, with precipitation levels 25 to 50 percent less than the long-term average. This was not the region's most severe drought if measured by total precipitation, but it was unusual in combining low precipitation with abnormally hot temperatures. Much of this region is covered with piñon and juniper woodlands, a vegetation type that is between a forest and a shrubland. By 2003 a large fraction of the piñons in the region died, with mortality greater than 90 percent in some areas (Breshears et al. 2005). The main cause of death was infestation by the pine bark beetle (*Ips confusus*), which often successfully attacks trees weakened by other stresses. The consequence of this mortality is a major change in the ecosystem structure and function over a large area. We do not, in general, know the thresholds for this kind of major change before we see them occur. It is possible that many ecosystems may be subject to dramatic changes at conditions only slightly outside the observed range, especially when they are subject to many interacting stresses.

Alaska and the Arctic

The Arctic is warming about twice as rapidly as the rest of the planet. Warming at high latitudes causes the sea ice and seasonal snow cover to melt more rapidly (Figure 11), converting white reflective surfaces to darker ocean water or vegetation, respectively. These dark surfaces absorb more solar radiation, transferring the heat to the air, leading to higher air temperatures and more rapid melting—a feedback loop that exposes arctic marine and terrestrial ecosystems to much higher rates of warming than the rest of the planet (Chapin et al. 2005). Permafrost, the permanently frozen ground characteristic of cold regions, contains about as much carbon as the global atmosphere (Zimov et al. 2006). As high-latitude regions warm, the thawing soil will release much of this carbon to the atmosphere, which in turn will cause more warming, in another continuous feedback loop. Exactly how fast this will happen is not known. In addition, as permafrost thaws and ice volume is lost, the ground subsides unevenly, forming small thaw ponds. The organic matter that decomposes in the airless sediments of these ponds produces methane, an even more powerful greenhouse gas than CO_2 (Walter et al. 2006). These powerful feedbacks to climate warming from permafrost thaw and many other finer-scale physical processes are not currently incorporated in global models. Uncertainty about the impact of these regional phenomena suggests that climate change could occur even more rapidly than models currently project.

Climate change is also affecting the way humans interact with arctic ecosystems. Shorter winters mean that ice roads used for oil exploration may no longer be practical. Alternative construction methods that lay gravel roads over sensitive tundra have much greater ecological impact on tundra than the ice roads and also affect the streams and rivers from which the gravel is taken. For example, female caribou and their calves avoid roads, thereby losing access to grazing areas, and the harvesting of stream gravels disrupt an important spawning habitat for fish.

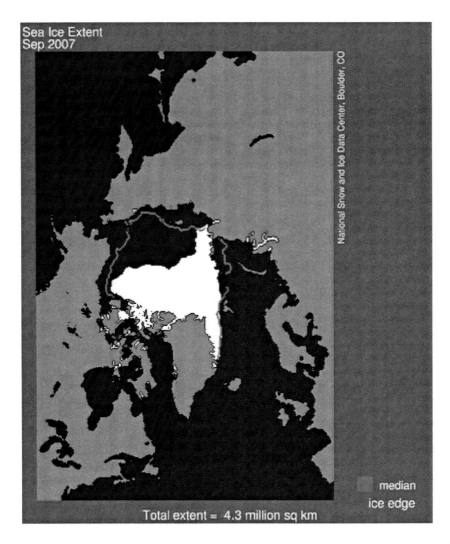

FIGURE 11 Average arctic sea ice area for the month of September 2007 (in white) and the average from 1979 to 2000 (pink outline). SOURCE: National Snow and Ice Data Center.

Changes in terrestrial vegetation and effects on the arctic food chain
Arctic warming has important consequences for the mixture of plants on the land. Under experimental conditions it has been shown that warming of tundra causes increased shrub growth. This vegetation change has also been observed in the field, based on repeat aerial photography, long-term field measurements, observations by indigenous people, and increased vegetation cover observed from satellites (Figure 12) (Chapin et al. 2005; Sturm et al. 2005; Goetz et al. 2005). If current trends continue, the future will see forests growing in areas previously dominated by shrubs and shrubs taking over in areas that used to hold rushes and sedges.

The expansion of shrub habitat is another example of a feedback loop. Because the shrubs are taller than the vegetation they replace, they tend to trap snow, preventing it from blowing away or turning directly from snow into water vapor. This increases the availability of meltwater in the spring, and the thicker blanket of snow insulates the soil, keeping it warmer over much of the winter. The arctic microbes respond dramatically to these warmer conditions, increasing the processing of soil organic matter and making more useable nitrogen. In tundra the shrubs grow faster in response to added nitrogen than do other species (Chapin et al. 2005), thus adding to their capacity to trap snow and further warm the soil (Sturm et al. 2005).

The expansion of tundra shrubs also has important ecological and social implications. The entire arctic ecosystem depends on caribou, including bears, wolves, and a range of carrion feeders, whose populations shift with the abundance of caribou. Caribou are also perhaps the most important terrestrial subsistence resource for indigenous peoples across the circumpolar Arctic. Lichens, an important winter food for caribou, are among the species that are crowded out by increased shrub growth (Cornelissen et al. 2001). The deeper snow around shrubs also makes it harder for caribou to reach the lichens beneath. And over the longer term, warming-induced increases in wildfire place an additional stress on this ecosystem because lichens recover from wildfire much more slowly than shrubs (Rupp et al. 2006).

A warmer climate may help caribou in the summer. Warmer summer temperatures tend to increase food availability and, as a consequence, survival of calves. But these advantages are countered by more frequent thaw events in winter, which tend to produce a layer of ice on top of the snow, making it difficult for caribou to reach the underlying foliage. Herd sizes have been observed to decrease during periods of frequent icing (Griffith et al. 2002).

FIGURE 12 Two photographs of the Ayiyak River in Alaska (68° 53'N, 152° 31'W), taken 50 years apart, showing larger individual shrubs, denser shrub patches, and expansion of shrubs into areas that were previously shrub-free. SOURCE: Sturm et al. 2001.

Ice-dependent animals: Walrus and polar bear
Climate change is having a major impact on the extent of sea ice, and therefore on the animals that depend on it, including walruses and polar bears. Walruses, for example, use ice floes as nursing platforms and as a home base from which they dive to feed on clams and other bottom-dwellers (Ray et al. 2006). Each spring walruses follow their sea-ice perches northward as ice floes in southern latitudes melt. Because of climate change, the range of year-round sea ice is shrinking and the walruses must move farther northward each year (Krupnik and Bogoslovskaya 1999; Grebmeier et al. 2006). In 2007 the sea ice moved beyond the edge of the continental shelf, where the water becomes too deep for the walruses to feed. For the first time in recorded history several thousand walruses—seeking an alternate place to rest between feeding excursions—set up camp along the beaches near the village of Wainwright, Alaska. Over time, such a dense aggregation of animals in a single location could deplete bottom food resources along the coast. This dense aggregation of animals also crushed many calves as adults moved to and from the ocean to feed (Metcalf and Robards 2008). Polar bears rely on sea ice for hunting; when the sea is covered with ice, bears can wait at openings in the ice for their favorite prey (ringed seals) to surface for air. In the open sea, seals can surface anywhere and the polar bears cannot catch them (Laidre et al. 2008).

Warmer waters and declines in sea ice (Figure 11) in the Bering Sea between Alaska and Russia are causing massive ecological changes (Grebmeier et al. 2006). Fish species are shifting northward in response to warmer temperatures and greater algal production in the water column. This reduces the organic matter that falls to the seafloor as sediments, reducing the productivity of the seafloor ecosystem on which walrus, crabs, and other species feed. As sea ice continues to retreat this entire ice-dependent ecosystem, including coastal indigenous communities that depend on marine mammals both culturally and nutritionally, will be substantially restructured (Grebmeier et al. 2006).

Lessons from the Distant Past

Much of what we know about the current ecological impacts of climate change comes from long-term observations and experiments. Our ability to predict future ecological impacts of climate change, however, largely stems from what we know about the effects of past climate changes. Although climate changes over recent geological history were generally more modest and slower than the changes we currently face, studying the geologic record of climate and ecosystem change provides a valuable way to understand how large-scale climate change affects natural vegetation and ecosystems.

Climatic records spanning the last 50,000 or so years can be generated from sediments, tree-rings, cave formations, corals, ice cores, and many other natural recorders of climate. These climate records can be dated and compared with vegetation and ecosystem records like those based on fossil pollen, plants, animals, and other organisms. These physical records allow scientists to understand what ecosystems were like in the past and together make up what is called the paleoecological record.

Past ecological responses to climate change

Thousands of years ago

The last 20,000 years on Earth have seen the demise of the last ice age, along with a global warming of 4-7°C (7.2-12.6° F) into the current interglacial period, all driven by subtle changes in the orbit of Earth (Jansen et al. 2007). This warming caused significant and widespread changes in dominant plant types as the ecosystems of the continent reorganized in response to climate warming and associated changes in the water cycle (Overpeck et al. 2003). This millennia-long climate shift caused some plant and animal species to become rare on the landscape and restricted to just a few isolated locations, a situation that can put species at greater risk of extinction. Some vegetation communities shifted to locations with more favorable conditions, others became extinct, and new communities emerged.

The paleoecological record of the last 20,000 years makes clear that each plant species adapts to large-scale climate change in its own way, and that the climate conditions needed for any individual species' survival and reproduction can move hundreds of kilometers or more across the landscape in response to climate change (Jackson and Overpeck, 2000). In extreme cases whole biomes (assemblages of plant species) can go extinct. For example, around 12,000 years ago, much of the U.S. Midwest was covered by a mixed forest of spruce and hardwoods unlike anything that can be found there today. This major vegetation biome went extinct as summers warmed and ice sheets retreated about 10,000 years ago. Of course, the converse is also true: when novel new climates emerge, ecosystems with entirely new mixtures of species may arise.

Interestingly, the long-term climate changes that occurred in the 20,000 years before the Industrial Revolution are known to have caused only one North American plant species—a type of spruce tree—to go extinct in the last 20,000 years (Jackson and Weng 1999). In contrast, many species of North American mammal—most notably Pleistocene megafauna such as wooly mammoths and mastodons—are known to have gone extinct over the last 20,000 years. Although debate continues on the exact cause of these extinctions, it appears most likely that climate change caused major reductions in the favored habitats of these animals, and that human hunting acted as a compounding stress (Koch and Barnosky 2006). This highlights the potentially deadly challenge that rapid climate change (even faster than over the last 20,000 years), combined with human activities (now more varied and pervasive than ever in the past), could pose to North American biodiversity in the future.

Millions of years ago

Another important lesson emerges from the Paleocene-Eocene Thermal Maximum, approximately 55 million years ago. At this time rapid and large increases in atmospheric carbon dioxide caused an abrupt, sustained global warming of over 5°C (9°F), and also acidification of the world's oceans—similar to what we currently face. Ocean chemistry took over 100,000 years to return to a less acidic state, and in the meantime many marine animals became extinct. Ecological impacts on land were also substantial, apparently including the appearance of the first primates on Earth, although the details of cause and effect remain uncertain (Jansen et al. 2007).

The global warming that occurred after the last ice age was large, particularly with respect to the magnitude of global warming over the last 100 years, but it was also likely at least 10 times slower than what could happen in the future. Such rapid changes in conditions place greater stress on ecosystems, since not all of the individual species that make up the ecosystem will be able to adapt or migrate at the same speed. In addition to unprecedented rates of warming, species now face human-caused fragmentation of the landscape and other barriers to migration, invasive species, groundwater and stream flow reductions, pollution, and other pervasive human influences that will inhibit the natural ability of ecosystems to adjust to climate change.

Impacts of Future Climate Changes

Uncertainties in predictions

The record of climate-induced change over the last million years indicates that human-caused climate change, if not slowed significantly, will have a major landscape-transforming impact across most of North America and its coastal ocean in the next 100 years and beyond. The lessons from the recent and distant past allow us to picture the likely impacts of future climate changes, but the picture is incomplete for several reasons. First, future climate changes will be unprecedented in many respects. Depending on human actions climate change over the next century may produce a temperature change as large as the difference between full glacial and full interglacial conditions, and with temperatures warmer than Earth has experienced in many millions of years. Second, if "business as usual" practices continue, climate change in coming decades will be exceptionally rapid, much more rapid on a sustained global basis than the transitions into and out of past ice ages. The only global-scale climate changes in Earth's history that have happened more rapidly are probably those associated with major cataclysms, such as meteor impacts. Third, climate change will occur in a setting where human actions have fundamentally altered terrestrial, aquatic, and marine ecosystems. Land use for farming and forestry has disrupted migration routes for some plants and animals, while improving them for others. Coastal ecosystems are increasingly squeezed between rising oceans and extensive human development along the coasts. Many rivers are dammed, diverted, or polluted. Wild stocks of fish are in many cases seriously depleted by overfishing or by changes in coastal and river habitats. Finally, human actions are very effectively facilitating species movements, both intentionally and unintentionally, making it possible for species that are good at moving to spread around the world, often eliminating native species in their path.

Making decisions in spite of uncertainty

Although evidence from the recent and distant past is incomplete, we can draw some important lessons. Perhaps the most important is that ecosystem responses to climate change, especially with interacting stresses, are extremely complex. Interactions that were unimportant in one setting may become critical in another. Healthy populations may be ravaged by pathogens that are newly at home in a formerly hostile or resistant environment. And some rare species may surprise us with their tenacity. Strategies for managing ecosystems in the future will need to pay special attention to uncertainty—making the best decisions based on available information and implementing decisions in a way that makes them adjustable as additional information becomes available.

Future climate change will affect many aspects of ecosystem composition, structure, and functioning. Some of these will have profound influences on ecosystem services. Others will have effects on the integrity of ecosystems and on their resilience (their ability to cope with future changes). Among all the possible impacts of climate change on ecosystems, the most permanent is extinction. Once a species is lost, it cannot be replaced. Everything that was unique about that species, perhaps its interactions with other species, perhaps its ability to deal with particular kinds of stresses, or perhaps its unique appearance or behavior, is lost forever. When we step back and probe the likely future consequences of human actions in causing climate change, increased extinctions are one of the key impacts. So far the number of known extinctions as a result of climate change is small, but quite a high number of species are currently considered functionally extinct, in other words, they are at risk of going extinct as the climate warms unless we directly intervene (Thomas et al. 2004). For example, species currently living at the top of mountains have no place else to go and will likely become extinct unless we capture them, move them to a more hospitable habitat and monitor them to make sure they survive in the new habitat (Hoegh-Guldberg et al. 2008). Such a response would require money, people power and political will. If a warming of 2-3°C (3.6-5.4° F) occurs, the Intergovernmental Panel on Climate Change has estimated that about 20 to 30 percent of studied species could risk extinction in the next 100 years. (Fischlin et al. 2007). Given that there are approximately 1.7 million identified species on the globe, an estimated 300,000 to 600,000 species could be committed to extinction primarily due to human activities. An important reason why climate change is expected to have such a major impact on biodiversity is that in most ecosystems climate change is occurring in the context of ongoing pressures from a range of other important factors, including loss of habitat from human land use, overfishing, fertilizer and pesticide runoff, and the encroachment of invasive species (Sala et al. 2000). Indeed, we seem to be standing at the brink of a mass-extinction event, precipitated by the behavior of one species—*Homo sapiens*.

What should we do about these trends?
Climate change is undoubtedly one of the defining environmental and development issues of the 21st century. Never before have humans had the numbers and the technology to dramatically alter the climate of Earth at the global scale. Decisions about climate change over the coming decades will likely reverberate through centuries.

This document is not intended to make policy recommendations. Rather, it is focused on characterizing some of the changes to ecosystems that have already occurred and that are likely to occur in the future, with different levels of climate change. There is no question that the impacts of climate change on ecosystems become increasingly profound as the magnitude and rate of climate change increases and that disruptions to ecosystem services, including potentially irreplaceable services from biological diversity, become more severe.

The challenge is finding a set of policies, practices, and standards of behavior that provide long-term economic opportunities and improved quality of life around the world while maintaining a sustainable climate and viable ecosystems. Some authoritative recent analyses (Stern 2007; IPCC 2007c) conclude that on economic grounds alone, the world should invest in curtailing the amount of climate change that occurs and in adapting to the changes that cannot be avoided. The appropriate level of these investments and the way they are financed and structured are relevant questions for a wide-ranging discussion among all members of society—in communities, businesses, places of worship, schools, and families. Some of the issues are quite technical and can only be effectively addressed at the level of governments. This includes decisions like whether and how best to impose a price on carbon emissions to the atmosphere or the kinds of technological alternatives to fossil fuel energy to receive government subsidies. Other decisions can be best addressed at the individual or family level. Each time a car, home appliance, or light bulb is purchased, a decision is made that has a small influence on the change in climate being driven by human-caused greenhouse gas emissions. But many small decisions, made by billions of people, can combine to have very large effects.

As has been illustrated throughout this report, climate change is not the only stress on our ecosystems. So another way that society can help reduce the negative ecological impacts of climate change is by creating conditions that make it easier for species in ecosystems to adapt. For example, the impacts of climate change on natural systems will be less harsh if other stresses on ecosystems that are in fact under human control are reduced. Ocean ecosystems could be strengthened by eliminating overfishing, guarding against invasive species, and reducing nutrient runoff. Ocean ecosystems could also be helped by protecting as much habitat and biodiversity as possible in a fashion that is designed to allow movement of species, for example, with networks of marine reserves where no fishing is allowed). Comparable protection on land would include preserves and parks connected by corridors. Carefully considered approaches to and investment in conservation, sustainable agricultural practices, pollution reduction, and water management can all work together to help ecosystems withstand the impact of a changing climate and maintain critical ecosystem services.

The climate challenge is big and complex. It is unlikely that it can be solved with any single strategy or by the people of any single country. But very likely it can be abated with the dedicated efforts of millions of people, working hard on diverse strategies, from many different angles.

REFERENCES AND SUGGESTIONS FOR FURTHER READING

ACIA (Arctic Climate Impact Assessment). 2004. Impacts of a Warming Arctic: Arctic Climate Impact Assessment. Cambridge, UK: Cambridge University Press.

ACIA. 2005. Arctic Climate Impact Assessment: Scientific Report. Cambridge, UK: Cambridge University Press.

Alley, R. B., P. U. Clark, P. Huybrechts, and I. Joughin. 2005. Ice-sheet and sea-level changes. Science 310:456-460.

Alvarez, W., F. Asaro, and A. Montanari. 1990. Iridium profile for 10 million years across the cretacious-tertiary boundary at Gubbio (Italy). Science 250:1700-1702.

Anderson, A., E. Allen, M. Dodero, T. Longcore, D. D. Murphy, C. Parmesan, G. Pratt, and M. C. Singer. 2001. Quino Checkerspot Butterfly (Euphydryas editha quino) Recovery Plan, Portland, Ore.: U.S. Fish and Wildlife Service.

Backlund, P., D. Schimel, A. Janetos, J. Hatfield, M. G. Ryan, S. R. Archer, and D. Lettenmaier. 2008. Introduction. IN: The effects of climate change on agriculture, land resources, water resources, and biodiversity in the United States: U.S. Climate Change Science Program, Synthesis and Assessment Product 4.3. ed. M. Walsh. pp. 11-20. Washington, D.C.: USDA.

Barnett, T. P., D. W. Pierce, H. G. Hidalgo, C. Bonfils, B. D. Santer, T. Das, G. Bala, A. W. Wood, T. Nozawa, and A. A. Mirin. 2008. Human-induced changes in the hydrology of the western United States. Science 319:1080.

Barth, J. A., B. A. Menge, J. Lubchenco, F. Chan, J. M. Bane, A. R. Kirincich, M. A. McManus, K. J. Nielsen, S. D. Pierce, and L. Washburn. 2007. Delayed upwelling alters nearshore coastal ocean ecosystems in the northern California current. PNAS 104(10):3719-3724.

Beaubien, E. G., and H. J. Freeland. 2000. Spring phenology trends in Alberta, Canada: Links to ocean temperature. International Journal of Biometeorology 44:53-59.

Beever, E. A., P. F. Brussard, and J. Berger. 2003. Patterns of apparent extirpation among isolated populations of pikas (Ochotona princeps) in the Great Basin. Journal of Mammalogy 84:37-54.

Bolin, B. 2007. A History of the Science and Politics of Climate Change: The Role of the Intergovernmental Panel on Climate Change. Cambridge, UK: Cambridge University Press.

Bomhard, M. B., D. M. Richardson, J. S. Donaldson, G. O. Hughes, G. F. Midgley, D. C. Raimondo, A. G. Rebelo, M. Rouget, and W. Thuiller. 2005. Potential impacts of future land use and climate change on the Red List status of the Cape Floristic Region, South Africa. Global Change Biology 11:1452-1468.

Both, C., S. Bouwhuis, C. M. Lessells, and M. E. Visser. 2006. Climate change and population declines in a long-distance migratory bird. Nature 441:81-83.

Breshears, D. D., N. S. Cobb, P. M. Rich, K. P. Price, C. D. Allen, R. G. Balice, W. H. Romme, J. H. Kastens, M. L. Floyd, J. Belnap, J. J. Anderson, O. B. Myers, and C. W. Meyer. 2005. Regional vegetation die-off in response to global-change-type drought. PNAS 102:15144-15148.

Brooks, M. L., and J. R. Matchett. 2006. Spatial and temporal patterns of wildfires in the Mojave Desert, 1980-2004. Journal of Arid Environments 67:148-164.

Caldeira, K., and M. E. Wickett. 2003. Anthropogenic carbon and ocean pH. Nature 425:365-365.

Cao, L., K. Caldeira, and A. K. Jain. 2007. Effects of carbon dioxide and climate change on ocean acidification and carbonate mineral saturation. Geophysical Research Letters 34:5.

Chan, F., J. A. Barth, J. Lubchenco, A. Kirincich, H. Weeks, W. T. Peterson, and B. A. Menge. 2008. Emergence of anoxia in the California current large marine ecosystem. Science 319:920.

Chapin, F. S., III, B. H. Walker, R. J. Hobbs, D. U. Hooper, J. H. Lawton, O. E. Sala, and D. Tilman. 1997. Biotic control over the functioning of ecosystems. Science 277:500-503.

Chapin, F. S., III, M. Sturm, M. C. Serreze, J. P. McFadden, J. R. Key, A. H. Lloyd, A. D. McGuire, T. S. Rupp, A. H. Lynch, J. P. Schimel, J. Beringer, W. L. Chapman, H. E. Epstein, E. S. Euskirchen, L. D. Hinzman, G. Jia, C. L. Ping, K. D. Tape, C. D. C. Thompson, D. A. Walker, and J. M. Welker. 2005. Role of land-surface changes in Arctic summer warming. Science 310:657-660.

Coope, G.R. 1995. Insect faunas in ice age environments: why so little extinction? IN: Extinction Rates, eds. J. H. Lawton, and R. M. May, pp. 55-74. Oxford, U.K.: Oxford University Press.

Coope, G.R. 1994. The response of insect faunas to glacial-interglacial climatic fluctuations. Philosophical Transactions of the Royal Society of London B 344:19-26.

Cornelissen, J. H. C., T. V. Callaghan, J. M. Alatalo, A. Michelsen, E. Graglia, A. E. Hartley, D. S. Hik, S. E. Hobbie, M. C. Press, C. H. Robinson, G. H. R. Henry, G. R. Shaver, G. K. Phoenix, D. Gwynn Jones, S. Jonasson, F. S. Chapin III, U. Molau, C. Neill, J. A. Lee, J. M. Melillo, B. Sveinbjornsson, and R. Aerts. 2001. Global change and arctic ecosystems: Is lichen decline a function of increases in vascular plant biomass? Journal of Ecology 89:984-994.

Covich, A. P., S. C. Fritz, P. J. Lamb, R. D. Marzolf, W. J. Matthews, K. A. Poiani, E. E. Prepas, M. B. Richman, and T. C. Winter. 1997. Potential effects of climate change on aquatic ecosystems of the Great Plains of North America. Hydrological Processes 11(8):993-1021.

Croxall, J. P., P. N. Trathan, and E. J. Murphy. 2002. Environmental change and antarctic seabird populations. Science 297:1510-1514.

Crozier, L. G., and R. W. Zabel. 2006. Climate impacts at multiple scales: Evidence for differential population responses in juvenile chinook salmon. Journal of Animal Ecology 75(5): 1100-1109.

Dai, A., K. E. Trenberth, and T. Qian. 2004. A global dataset of Palmer Drought Severity Index for 1870-2002: Relationship with soil moisture and effects of surface warming. Journal of Hydrology 5:117-1129.

Daily, G. C., editor. 1997. Nature's Services. Washington D.C.: Island Press.

Davis, M. B., and R. G. Shaw. 2001. Range shifts and adaptive responses to quaternary climate change. Science 292:673-679.

Day, J. W., Jr., D. F. Boesch, E. J. Clairain, G. P. Kemp, S. B. Laska, W. J. Mitsch, K. Orth, H. Mashriqui, D. R. Reed, L. Shabman, C. A. Simenstad, B. J. Streever, R. R. Twilley, C. C. Watson, J. T. Wells, and D. F. Whigham. 2007. Restoration of the Mississippi Delta: Lessons from hurricanes Katrina and Rita. Science 315:1679-1684.

Derocher, A. E., N. J. Lunn, and I. Stirling. 2004. Polar bears in a warming climate. Integrative and Comparative Biology 44:163-176.

Diaz, R. J., and R. Rosenberg. 2008. Spreading dead zones and consequences for marine ecosystems. Science 321(5891):926-929.

Doney, S. C., V. J. Fabry, R. A. Feely, and J. A. Kleypas. Forthcoming: Annual Review of Marine Sciences. Available at http://arjournals.annualreviews.org/toc/marine/forthcoming.

Ducklow, H. W., K. Baker, D. G. Martinson, L. B. Quetin, R. M. Ross, R. C. Smith, S. E. Stammerjohn, M. Vernet, and W. Fraser. 2007. Marine Pelagic Ecosystems: the West Antarctic Peninsula. IN: Antarctic Ecology: From Genes to Ecosystems, eds. A. Rogers, E. Murphy, and A. C. Clarke. Philosophical Transactions Royal Society B 362 (1477):67-94.

Edwards, M., and A. J. Richardson. 2004. Impact of climate change on marine pelagic phenology and trophic mismatch. Nature 430:881-884.

Ellis, E. C., and N. Ramankutty. 2008. Putting people in the map: Anthropogenic biomes of the world. Frontiers in Ecology and the Environment 6.

Emanuel, K. 2005. Increasing destructiveness of tropical cyclones over the past 30 years. Nature 436:686-688.

Emanuel, K., R. Sundararajan, and J. Williams. 2008. Hurricanes and global warming—Results from downscaling IPCC AR4 simulations. Bulletin of the American Meteorological Society 89:347-367.

Esque, T. C., C. R. Schwalbe, D. F. Haines, and W. L. Halvorson. 2004. Saguaros under siege: Invasive species and fire. Desert Plants 20:49-55.

Fabry, V. J., B. A. Seibel, R. A. Feely, and J. C. Orr. 2008. Impacts of ocean acidification on marine fauna and ecosystem processes. ICES Journal of Marine Sciences 65:414.

Ferguson, S. H., I. Stirling, and P. McLoughlin. 2005. Climate change and ringed seal (Phoca hispida) recruitment in western Hudson Bay. Marine Mammal Science 21:121-135.

Fetterer, F., K. Knowles, W. Meier, and M. Savoie. 2002, updated 2008. Sea Ice Index. Digital Media. Boulder, Colo.: National Snow and Ice Data Center.

Field, C. B., D. B. Lobell, H. A. Peters, and N. R. Chiariello. 2007a. Feedbacks of terrestrial ecosystems to climate change. Annual Review of Environment and Resources 32:1-29.

Field, C. B., L. D. Mortsch, M. Brklacich, D. L. Forbes, P. Kovacs, J. A. Patz, S. W. Running, and M. J. Scott. 2007b. North America. IN: Climate Change 2007: Impacts, Adaptation and Vulnerability. Contribution of Working Group II to the Fourth Assessment Report of the Intergovernmental Panel on Climate Change. eds. M. L. Parry, O. F. Canziani, J. P. Palutikof, P. J. van der Linden, and C. E. Hanson, pp. 617-652. Cambridge U.K.: Cambridge University Press.

Fischlin, A., G. F. Midgley, J. T. Price, R. Leemans, B. Gopal, C. Turley, M. D. A. Rounsevell, O. P. Dube, J. Tarazona, and A. A. Velichko. 2007. Ecosystems, their properties, goods, and services. IN: Climate Change 2007: Impacts, Adaptation and Vulnerability. Contribution of Working Group II to the Fourth Assessment Report of the Intergovernmental Panel on Climate Change. eds. M. L. Parry, O. F. Canziani, J. P. Palutikof, P. J. van der Linden, and C.E. Hanson, pp. 211-272. Cambridge U.K.: Cambridge University Press.

Fogarty, M., L. Incze, K. Hayhoe, D. Mountain and J. Manning. 2008. Potential climate impacts on Atlantic cod (Gadus morhua) off the northeastern USA. Mitigation and Adaptation Strategies to Global Change 13:453-466.

Ford, S. E. 1996. Range extension by the oyster parasite Perkinsus marinus into the northeastern United States: Response to climate change? Journal of Shellfish Research 15:45-56.

Forster, P., V. Ramaswamy, P. Artaxo, T. Berntsen, R. Betts, D. W. Fahey, J. Haywood, J. Lean, D. C. Lowe, G. Myhre, J. Nganga, R. Prinn, G. Raga, M. Schulz and R. Van Dorland. 2007. Changes in atmospheric constituents and in radiative forcing. IN: Climate Change 2007: The Physical Science Basis. Contribution of Working Group I to the Fourth Assessment Report of the Intergovernmental Panel on Climate Change eds. S. Solomon, D. Qin, M. Manning, Z. Chen, M. Marquis, K. B. Averyt, M. Tignor, and H. L. Miller. Cambridge, U.K.: Cambridge University Press.

Frumhoff, P. C., J. J. McCarthy, J. M. Melillo, S. C. Moser, and D. J. Wuebbles. 2007. Confronting Climate Change in the U.S. Northeast: Science, Impacts, and Solutions. Synthesis report of the Northeast Climate Impacts Assessment (NECIA). Cambridge, Mass.: Union of Concerned Scientists.

Glenn, R. P., and T. L. Pugh. 2006. Epizootic shell disease in American lobster (Homarus americanus) in Massachusetts coastal waters: Interactions of temperature, maturity, and intermolt duration. Journal of Crustacean Biology 26(4):639-645.

Goetz, S. J., A. G. Bunn, G. A. Fiske, and R. A. Houghton. 2005. Satellite-observed photosynthetic trends across boreal North America associated with climate and fire disturbance. PNAS 102:13521-13525.

Grebmeier, J. M., J. E. Overland, S. E. Moore, E. V. Farley, E. C. Carmack, L. W. Cooper, K. E. Frey, J. H. Helle, F. A. McLaughlin, and S. L. McNutt. 2006. A major ecosystem shift in the northern Bering Sea. Science 311:1461-1464.

Gregory, J., and P. Huybrechts. 2006. Ice-sheet contributions to future sea-level change. Philosophical Transactions of the Royal Society of London A 364:1709-1731.

Griffith, B., D. C. Douglas, N. E. Walsh, D. D. Young, T. R. McCabe, D. E. Russell, R. G. White, R. D. Cameron, and K. R. Whitten. 2002. The porcupine caribou herd. IN: Arctic Refuge coastal plain terrestrial wildlife research summaries, eds. D. C. Douglas, P. E. Reynolds, and E. B. Rhode. pp. 8-37. U. S. Geological Survey, Biological Resources Division, Biological Science Report USGS/BRD/BSR-2002-0001.

Gunderson, L. A., and L. Pritchard Jr. 2002. Resilience and the Behaviour of Large-Scale Systems. Washington, D.C.: Island Press.

Hatfield, J. L., K. J. Boote, B. A. Kimball, D. W. Wolfe, D. R. Ort, C. R. Izaurralde, A. M. Thomson, J. A. Morgan, H. W. Polley, P. A. Fay, T. L. Mader, and G. L. Hahn. 2008. Agriculture. IN: The effects of climate change on agriculture, land resources, water resources, and biodiversity in the United States: U.S. Climate Change Science Program, Synthesis and Assessment Product 4.3. ed. M. Walsh. pp. 21-74. Washington, D.C.: USDA.

Hayhoe, K., D. Cayan, C. B. Field, P. C. Frumhoff, E. P. Maurer, N. L. Miller, S. C. Moser, S. H. Schneider, K. N. Cahill, E. E. Cleland, L. Dale, R. Drapek, R. M. Hanemann, L. S. Kalkstein, J. Lenihan, C. K. Lunch, R. P. Neilson, S. C. Sheridan, and J. H. Verville. 2004. Emissions pathways, climate change, and impacts on California. PNAS 101(34): 12422-12427.

Hill, G. E., R. R. Sargent, M. B. Sargent. 1998. Recent change in the winter distribution of Rufous Hummingbirds. The Auk 115:240-245.

Hinzman, L. D., N. D. Bettez, W. R. Bolton, F. S. Chapin, M. B. Dyurgerov, C. L. Fastie, B. Griffith, R. D. Hollister, A. Hope, H. P. Huntington, A. M. Jensen, G. J. Jia, T. Jorgenson, D. L. Kane, D. R. Klein, G. Kofinas, A. H. Lynch, A. H. Lloyd, A. D. McGuire, F. E. Nelson, W. C. Oechel, T. E. Osterkamp, C. H. Racine, V. E.

Romanovsky, R. S. Stone, D. A. Stow, M. Sturm, C. E. Tweedie, G. L. Vourlitis, M. D. Walker, D. A. Walker, P. J. Webber, J. M. Welker, K. Winker, and K. Yoshikawa. 2005. Evidence and implications of recent climate change in northern Alaska and other arctic regions. Climatic Change 72:251-298.

Hoegh-Guldberg, O., P. J. Mumby, A. J. Hooten, R. S. Steneck, P. Greenfield, E. Gomez, C. D. Harvell, P. F. Sale, A. J. Edwards, K. Caldeira, N. Knowlton, C. M. Eakin, R. Iglesias-Prieto, N. Muthiga, R. H. Bradbury, A. Dubi, and M. E. Hatziolos. 2007. Coral reefs under rapid climate change and ocean acidification. Science 318:1737-1742.

Hoegh-Guldberg, O., Hughes, L, McIntyre, S. L., Lindenmayer, D. B., Parmesan, C., Possingham, H. P., Thomas, C. D. 2008. Assisted colonization and rapid climate change. Science 321:345-346.

Hofmann, G. E., M. J. O'Donnell, M. A. Sewell, L. M. Hammond, A. E. Todgham, and M. L. Zippay. 2008. Oral presenation. Does elevated CO_2 affect larval skeletal development in sea urchins?: Exploring the mechanisms with gene expression analysis and morphometrics. American Society of Limnology and Oceanography Ocean Sciences Meeting, Orlando, FL Vol. Eos Trans OS003, Abstract 3236.

Inouye, D. W., B. Barr, K. B. Armitage, and B. D. Inouye. 2000. Climate change is affecting altitudinal migrants and hibernating species. PNAS 97:1630-1633.

IPCC (Intergovernmental Panel on Climate Change). 2007a. Summary for policymakers. IN: Climate Change 2007: Impacts, Adaptation and Vulnerability. Contribution of Working Group II to the Fourth Assessment Report of the Intergovernmental Panel on Climate Change, eds. M. L. Parry, O. F. Canziani, J. P. Palutikof, P. J. v. d. Linden, and C. E. Hanson.Cambridge, U.K.: Cambridge University Press.

IPCC. 2007b. Summary for policymakers. IN: Climate Change 2007: The Physical Science Basis: Contribution of Working Group I to the Fourth Assessment Report of the Intergovernmental Panel on Climate Change, eds. S. Solomon, D. Qin, M. Manning, Z. Chen, M. Marquis, K. B. Averyt, M.Tignor, and H. L. Miller. pp. 1-21. Cambridge, U.K.: Cambridge University Press.

IPCC. 2007c. Summary for policymakers. IN: Climate Change 2007: Mitigation. Contribution of Working Group III to the Fourth Assessment Report of the Intergovernmental Panel on Climate Change. eds., B. Metz, O. R. Davidson, P. R. Bosch, R. Dave, and L. A. Meyer. pp. 1-23. Cambridge, U.K.: Cambridge University Press.

IPCC. 2007d. Climate Change 2007: Synthesis Report. Contribution of Working Groups I, II and III to the Fourth Assessment Report of the Intergovernmental Panel on Climate Change, eds. R. K. Pachauri and A. Reisinger. Geneva: IPCC.

Jackson, S. T., and J. T. Overpeck. 2000. Responses of plant populations and communities to environmental changes of the late Quaternary. IN: Deep Time: Paleobiology's Perspective. eds., D. H. Erwin and S. L. Wing. Paleobiology 26(Suppl. 4):194-220.

Jackson, S. T., and C. Weng. 1999. Late Quaternary extinction of a tree species in eastern North America. PNAS 96(24):13847-13852.

Jansen, E., J. Overpeck, K. R. Briffa, J. C. Duplessy, F. Joos, V. Masson-Delmotte, D. Olago, B. Otto-Bliesner, W. R. Peltier, S. Rahmstorf, R. Ramesh, D. Raynaud, D. Rind, O. Solomina, R. Villalba, and D. Zhang. 2007. Palaeoclimate. IN: Climate Change 2007: The Physical Science Basis. Contribution of Working Group I to the Fourth Assessment Report of the Intergovernmental Panel on Climate Change, eds. S. Solomon, D. Qin, M.

Manning, Z. Chen, M. Marquis, K. B. Averyt, M. Tignor, and H. L. Miller. Cambridge, U.K.: Cambridge University Press.

Johnson, T., J. Dozier, and J. Michaelsen. 1999. Climate change and Sierra Nevada snowpack. IN: Interactions between the cryosphere, climate and greenhouse gases, ed. Tranter, M. pp. 63-70. Wallingford, Oxfordshire, U.K.: IAHS Press.

Johnson, W. C., B. V. Millett, T. Gilmanov, R. A. Voldseth, G. R. Guntenspergen, and D. E. Naugle. 2005. Vulnerability of northern prairie wetlands to climate change. BioScience 55(10):863–872.

Jouzel, J., V. Masson-Delmotte, O. Cattani, G. Dreyfus, S. Falourd, G. Hoffmann, B. Minster, J. Nouet, J. M. Barnola, and J. Chappellaz. 2007. Orbital and millennial antarctic climate variability over the past 800,000 years. Science 317:793-793.

Karl, T. R., R. W. Knight, D. R. Easterling, and R. G. Quayle. 1996. Indices of climate change for the United States. Bulletin of the American Meteorological Society. 77:279-292.

Knowles, N. 2006. Trends in snowfall versus rainfall in the western United States. Journal of Climate 19(18):4545-4559.

Koch, P. L., and A. D. Barnosky. 2006. Late Quaternary extinctions: State of the debate. Annual Review of Ecology, Evolution, and Systematics 37:215-250.

Krupnik, I., and L. Bogoslovskaya. 1999. Old records, new stories: Ecosystem variability and subsistence hunting pressure in the Bering Strait area. Arctic Research of the United States 13:15-24.

Kunkel, K., P. Bromirski, H. Brooks, T. Cavasos, A. Douglas, D. Easterling, K. Emanuel, P. Groisman, G. Holland, T. Knutson, J. Kossin, P. Komar, D. Levinson, and R. Smith. 2008. Observed changes of weather and climate extremes. IN: Weather and Climate Extremes in a Changing Climate. Eds. Karl T. R., and G. R. Meehl. U.S. Climate Change Science Program, Synthesis Assessment Product 3.3, Chapter 2. Executive Office of the President.

Laidre, K. L., I. Stirling, L. F. Lowry, Ø. Wiig, M. P. Heide-Jørgensen, and S. H. Ferguson. 2008. Quantifying the sensitivity of arctic marine mammals to climate-induced habitat change. Ecological Applications 18:97-125.

Lobell, D. B., and G. P. Asner. 2003. Climate and management contributions to recent trends in US agricultural yields. Science 299:1032.

Logan, J. A., J. Regniere, and J. A. Powell. 2003. Assessing the impacts of global warming on forest pest dynamics. Frontiers in Ecology and the Environment 1(3):130-137.

Long, S. P., E. A. Ainsworth, A. D. B. Leakey, J. Nösberger, and D. R. Ort. 2006. Food for thought: Lower-than-expected crop yield stimulation with rising CO_2 concentrations. Science 312:1918-1921.

Matthews, H. D., and K. Caldeira. 2008. Stabilizing climate requires near-zero emissions. Geophysical Research Letters 35.

Meehl, G. A., T. F. Stocker, W. D. Collins, P. Friedlingstein, A. T. Gaye, J. M. Gregory, A. Kitoh, R. Knutti, J. M. Murphy, A. Noda, S. C. B. Raper, I. G. Watterson, A. J. Weaver and Z.-C. Zhao. 2007. Global Climate Projections. IN: Climate Change 2007: The Physical Science Basis. Contribution of Working Group I to the Fourth Assessment Report of the Intergovernmental Panel on Climate Change, eds. Solomon, S., D. Qin, M. Manning, Z. Chen, M. Marquis, K.B. Averyt, M. Tignor and H.L. Miller. Cambridge, U.K.: Cambridge University Press.

Metcalf, V., and M. D. Robards. 2008. Sustaining a healthy human-walrus relationship in a dynamic environment: Challenges for comanagement. Ecological Applications 18(2):S148-S156.

Millennium Ecosystem Assessment. 2005. Ecosystems and Human Well-being: Synthesis. Washington, D.C., Island Press.

Mohan, J. E., L. H. Ziska, W. H. Schlesinger, R. B. Thomas, R. C. Sicher, K. George, and J. S. Clark. 2006. Biomass and toxicity responses of poison ivy (Toxicodendron radicans) to elevated atmospheric CO_2. PNAS 103(24):9086-9089.

Mote, P. 2003. Trends in snow water equivalent in the Pacific Northwest and their climatic causes. Geophysical Research Letters 30(12):1601.

Mote, P., A. F. Hamlet, M. P. Clark, and D. P. Lettenmaier. 2005. Declining mountain snowpack in western North America. Bulletin of the American Mcteorological Society 86:39-49.

Mueter, F. J., and M. A. Litzow. 2008. Sea ice retreat alters the biogeography of the Bering Sea continental shelf. Ecological Applications 18(2):309-320.

National Agricultural Statistics Service. 2008. The Census of Agriculture. Available at http://www.agcensus.usda.gov/Publications/Historical_Publications/index.asp.

Nemani, R. R., M. A. White, D. R. Cayan, G. V. Jones, S. W. Running, J. C. Coughlan, and D. L. Peterson. 2001. Asymmetric warming over coastal California and its impact on the premium wine industry. Climate Research 19:25-34.

Nemani, R. R., C. D. Keeling, H. Hashimoto, W. M. Jolly, S. C. Piper, C. J. Tucker, R. B. Myneni, and S. W. Running. 2003. Climate-driven increases in global terrestrial net primary production from 1982 to 1999. Science 300:1560-1563.

NRC (National Research Council). 2008. Progress toward restoring the Everglades: The Second Biennial Review. Washington D.C.: The National Academies Press.

Opdam, P., and D. Wascher. 2004. Climate change meets habitat fragmentation: Linking landscape and biogeographical scale levels in research and conservation. Biological Conservation 117:285-297.

Orr, J. C., V. J. Fabry, O. Aumont, L. Bopp, S. C. Doney, R. A. Feely, A. Gnanadesikan, N. Gruber, A. Ishida, F. Joos, R. M. Key, K. Lindsay, E. Maier-Reimer1, R. Matear, P. Monfray, A. Mouchet, R. G. Najjar, G. K. Plattner, K. B. Rodgers, C. L. Sabine, J. L. Sarmiento, R. Schlitzer, R. D. Slater, I. J. Totterdell, M. F. Weirig, Y. Yamanaka, and A. Yool. 2005. Anthropogenic ocean acidification over the twenty-first century and its impact on calcifying organisms. Nature 437:681-686.

Overpeck, J. T., C. Whitlock, and B. Huntley. 2003. Terrestrial biosphere dynamics in the climate system: past and future. IN: Paleoclimate, Global Change and the Future (IGBP Synthesis Volume), eds. K. Alverson, R. Bradley and T. Pedersen. pp. 81-111. Berlin, Springer-Verlag.

Overpeck, J. T., B. L. Otto-Bliesner, G. H. Miller, D. R. Muhs, R. B. Alley, and J. T. Kiehl. 2006. Paleoclimatic evidence for future ice-sheet instability and rapid sea-level rise. Science 311:1747-1750.

Pandolfi, J. M., J. B. C. Jackson, N. Baron, R. H. Bradbury, H. M. Guzman, T. P. Hughes, C. V. Kappel, F. Micheli, J. C. Ogden, H. P. Possingham, and E. Sala. 2005. Are U.S. coral reefs on the slippery slope to slime? Science 307:1725-1728.

Parmesan, C. 1996. Climate and species' range. Nature 382:765-766.

Parmesan, C. 2006. Ecological and evolutionary responses to recent climate change. Annual Review of Ecology, Evolution, and Systematics 37:637-669.

Parmesan, C. 2007. Influence of species, latitudes and methodologies on estimates of phenological response to global warming. Global Change Biology 13:1860-1872.

Parmesan, C., and G. Yohe. 2003. A globally coherent fingerprint of climate change impacts across natural systems. Nature 421:37-42.

Paulson, D. R. 2001. Recent odonata records from southern Florida: Effects of global warming? International Journal of Odonatology 4:57-69.

Pauly, D., and V. Christensen. 1995. Primary production required to sustain global fisheries. Nature 374:255-257.

Pitelka, L. F., R. H. Gardner, J. Ash, S. Berry, H. Gitay, I. R. Noble, A. Saunders, R. H. W. Bradshaw, L. Brubaker, J. S. Clark, and M. B. Davis. 1997. Plant migration and climate change. American Scientist 85:464-473.

Raffa, K. F., B. H. Aukema, B. J. Bentz, A. L. Carroll, J. A. Hicke, M. G. Turner, and W. H. Romme. 2008. Cross-scale drivers of natural disturbances prone to anthropogenic amplification: The dynamics of bark beetle eruptions. BioScience 58(6):501-517.

Rahmstorf, S. 2007. A Semi-empirical approach to projecting future sea-level rise. Science 315(5810):368-370.

Ramanathan, V., M. V. Ramana, G. Roberts, D. Kim, C. Corrigan, C. Chung, and D. Winker. 2007. Warming trends in Asia amplified by brown cloud solar absorption. Nature 448:575-578.

Raupach, M. R., G. Marland, P. Ciais, C. Le Quere, J. G. Canadell, G. Klepper, and C. B. Field. 2007. Global and regional drivers of accelerating CO_2 emissions. PNAS 104:10288-10293.

Raven, J. K., Caldeira, H. Elderfield, O. Hoegh-Guldberg, P. Liss, U. Riebesell, J. Shepherd, C. Turley, and A. Watson. 2005. Ocean acidification due to increasing atmospheric carbon dioxide. London: The Royal Society.

Ray, G. C., J. McCormick-Ray, P. Berg, and H. E. Epstein. 2006. Pacific walrus: Benthic bioturbator of Beringia. Journal of Experimental Marine Biology and Ecology 330:403-419.

Rieman, B. E., D. Isaak, S. Adams, D. Horan, D. Nagel, C. Luce, and D. Myers. 2007. Anticipated climate warming effects on bull trout habitats and populations across the interior Columbia River basin. Transactions of the American Fisheries Society 136:1552–1565.

Root, T. L., and S. H. Schneider. 2002. Climate change: Overview and implications for wildlife. IN: Wildlife Responses to Climate Change: North American Case Studies, eds. S.H. Schneider, and T.L. Root. pp. 437. Washington, D.C.: Island Press.

Root, T. L., D. P. MacMynowski, M. D. Mastrandrea, and S. H. Schneider. 2005. Human-modified temperatures induce species changes: Joint attribution. PNAS 102:7465-7469.

Root, T. L., J. T. Price, K. R. Hall, S. H. Schneider, C. Rosenzweig, and J. A. Pounds. 2003. Fingerprints of global warming on animals and plants. Nature 421:57-60.

Rosenzweig, C., G. Casassa, D. J. Karoly, A. Imeson, C. Liu, A. Menzel, S. Rawlins, T. L. Root, B. Seguin, and P. Tryjanowski. 2007. Assessment of observed changes and responses in natural and managed systems IN: Climate Change 2007: Impacts, Adaptation and Vulnerability. Contribution of Working Group II to the Fourth Assessment Report of the Intergovernmental Panel on Climate Change, eds. M. L. Parry, O. F. Canziani, J. P. Palutikof, P. J. v. d. Linden, and C. E. Hanson. pp. 79-131. Cambridge, U.K.: Cambridge University Press.

Ross, M. S., P. L. Ruiz, G. J. Telesnicki, and J. F. Meeder. 2000. Estimating aboveground biomass and production in mangrove communities of Biscayne National Park, Florida. Wetlands Ecology and Management 9:27-37.

Rothrock, D. A., D. B. Percival, and M. Wensnahan. 2008. The decline in arctic sea-ice thickness: Separating the spatial, annual, and interannual variability in a quarter century of submarine data. Journal of Geophysical Research 113.

Ruddiman, W. F. 2007. The early anthropogenic hypothesis: Challenges and responses. Reviews of Geophysics 45:RG4001.

Rupp, T. S., M. Olson, J. Henkelman, L. Adams, B. Dale, K. Joly, W. Collins, and A. M. Starfield. 2006. Simulating the influence of a changing fire regime on caribou winter foraging habitat. Ecological Applications 16:1730-1743.

Sagarin, R. D., J. P. Barry, S. E. Gilman, and C. H. Baxter. 1999. Climate-related change in an intertidal community over short and long time scales. Ecological Monographs 69:465-490.

Sala, O. E., F. S. Chapin III, J. J. Armesto, E. Berlow, J. Bloomfield, R. Dirzo, E. Huber-Sanwald, L. F. Huenneke, R. B. Jackson, A. Kinzig, R. Leemans, D. M. Lodge, H. A. Mooney, M. Oesterheld, N. L. Poff, M. T. Sykes, B. H. Walker, M. Walker, and D. H. Wall. 2000. Global biodiversity scenarios for the year 2100. Science 287:1770-1774.

Schneider, S. H., and T. L. Root, eds. 2001. Wildlife Responses to Climate Change: North American Case Studies. Washington, D.C.: Island Press.

Schulze, E. D., and H. A. Mooney, eds. 1993. Biodiversity and Ecosystem Function. Berlin: Springer-Verlag.

Seager, R., M. Ting, I. Held, Y. Kushnir, J. Lu, G. Vecchi, H. P. Huang, N. Harnik, A. Leetmaa, N. C. Lau, C. Li, J. Velez, and N. Naik. 2007. Model projections of an imminent transition to a more arid climate in southwestern North America. Science 316(5828):1181.

Seimon, T. A., A. Seimon, P. Daszak, S. R. P. Halloys, L. M. Schloegel, C. A. Aguilar, P. Sowell, A. D. Hyatt, B. Konecky, J. E. Simmons. 2007. Upward range extension of Andean anurans and chytridiomycosis to extreme elevations in response to tropical deglaciation. Global Change Biology 13:288-299.

Serreze, M. C., M. M. Holland, and J. Stroeve. 2007. Perspectives on the Arctic's shrinking sea-ice cover. Science 315:1533-1536.

Singer, M. C. 1971. Ph.D. dissertation, Stanford University.

Singer, M. C. 1972. Complex components of habitat suitability within a butterfly colony. Science 173:75-77.

Singer M. C., P. R. Ehrlich. 1979. Population dynamics of the checkerspot butterfly Euphydryas editha. Fortschritte der Zoologie 25:53-60.

Singer M. C., and C. D. Thomas. 1996. Evolutionary responses of a butterfly metapopulation to human and climate-caused environmental variation. The American Naturalist 148:S9-39.

Smith, A. T. 1974. The distribution and dispersal of pikas: Influences of behavior and climate. Ecology 55:1368-1376.

Smith, J. R., P. Fong, and R. F. Ambrose. 2006. Dramatic declines in mussel bed community diversity: Response to climate change? Ecology 87(5):1153-1161.

Southward, A. J., O. Langmead, N. J. Hardman-Mountford, J. Aitken, G. T. Boalch, P. R. Dando, M. J. Genner, I. Joint, M. A. Kendall, N. C. Halliday, R. P. Harris, R. Leaper, N. Mieszkowska, R. D. Pingree, A. J. Richardson, D. W. Sims, T. Smith, A. A. Walne, and

S. J. Hawkins. 2005. Long-term oceanographic and ecological research in the western English Channel. Advances in Marine Biology 47:1-105.

Stern, N. 2007. The Economics of Climate Change: The Stern Review. Cambridge, U.K.: Cambridge University Press.

Stirling, I., N. J. Lunn, J. Iacozza. 1999. Long-term trends in the population ecology of polar bears in western Hudson Bay in relation to climatic change. Arctic 52:294-306.

Stramma, L., G. C. Johnson, J. Sprintall, and V. Mohrholz. 2008. Expanding oxygen-minimum zones in the tropical oceans. Science 320:655-658.

Sturm, M., C. Racine and K. Tape. 2001. Climate change: Increasing shrub abundance in the Arctic. Nature 411:546-547.

Sturm, M., J. Schimel, G. Michaelson, J. M. Welker, S. F. Oberbauer, G. E. Liston, J. Fahnestock, and V. E. Romanovsky. 2005. Winter biological processes could help convert arctic tundra to shrubland. BioScience 55:17-26.

Thomas, C. D., A. Cameron, R. E. Green, M. Bakkenes, L. J. Beaumont, Y. C. Collingham, B. F. N. Erasmus, M. F. d. Siqueira, A. Grainger, L. Hannah, L. Hughes, B. Huntley, A. S. v. Jaarsveld, G. F. Midgley, L. Miles, M. A. Ortega-Huerta, A. T. Peterson, O. L. Phillips, and S. E. Williams. 2004. Extinction risk from climate change. Nature 427:145-148.

Thuiller,W., S. Lavorel, M. B. Araujo, M. T. Sykes and I. C. Prentice. 2005. Climate change threats to plant diversity in Europe. PNAS 102:8245-8250.

Tilman, D., P. B. Reich, and J. M. H. Knops. 2006. Biodiversity and ecosystem stability in a decade-long grassland experiment. Nature 441:629-632.

Turner, M. G., W. H. Romme, R. H. Gardner, and W. W. Hargrove. 1997. Effects of fire size and pattern on early succession in Yellowstone National Park. Ecological Monographs 67:411-433.

Twilley, R. R., E. J. Barron, H. L. Gholz, M. A. Harwell, R. L. Miller, D. J. Reed, J. B. Rose, E. H. Siemann, R. G. Wetzel and R. J. Zimmerman. 2001. Confronting Climate Change in the Gulf Coast Region: Prospects for Sustaining Our Ecological Heritage. Washington, D.C.: Union of Concerned Scientists, Cambridge, Massachusetts, and Ecological Society of America.

UNEP (United Nations Environment Programme). 2007. In Dead Water: Disastrous merging of climate change with pollution, over-harvest, and infestations in the world's fishing grounds. A UNEP Rapid Response Assessment.

Walter, H. 1968. Die Vegetation der Erde in öko-physiologischer Betrachtung. Jena, Ger.: Fischer.

Walter, M. K., S. A. Zimov, J. P. Chanton, D. Verbyla, and F. S. Chapin III. 2006. Methane bubbling from Siberian thaw lakes as a positive feedback to climate warming. Nature 443:71-75.

Webster, P. J., G. J. Holland, J. A. Curry, H.-R. Chang. 2005. Changes in Tropical Cyclone Number, Duration, and Intensity in a Warming Environment. Science 309:1844-1846.

Westerling, A. L., H. G. Hidalgo, D. R. Cayan, and T. W. Swetnam. 2006. Warming and earlier spring increase western U.S. forest wildfire activity. Science 313:940-943.

White, M. A., N. S. Diffenbaugh, G. V. Jones, J. S. Pal, and F. Giorgi. 2006. Extreme heat reduces and shifts United States premium wine production in the 21st century. PNAS 103:11217-11222.

Williams, S. E., E. E. Bolitho and S. Fox. 2003. Climate change in Australian tropical rainforests: An impending environmental catastrophe. Proceedings of the Royal Society B 270:1887-1892.

Worm, B., E. B. Barbier, N. Beaumont, J. E. Duffy, C. Folke, B. S. Halpern, J. B. C. Jackson, H. K. Lotze, F. Micheli, and S. R. Palumbi. 2006. Impacts of biodiversity loss on ocean ecosystem services. Science 314:787-790.

Zimov, S. A., E. A. G. Schuur, and F. S. Chapin III. 2006. Climate Change: Permafrost and the global carbon budget. Science 312:1612.

Appendix A

The Committee on Ecological Impacts of Climate Change

Statement of Task

An ad-hoc committee will prepare a short (20-page) consensus, high-level overview report based on the recent Intergovernmental Panel on Climate Change Working Group II report and other NRC reports and published literature that provides a balanced summary of key examples of observed impacts of climate change on a variety of ecosystems. This report will form the basis of derivative products such as a public brochure and material that could be used as the basis of a class curriculum or public education.

Appendix B

Committee on Ecological Impacts of Climate Change
Member Biographies

Christopher B. Field – Chair
Carnegie Institution for Science

Christopher B. Field is the founding director of the Carnegie Institution's Department of Global Ecology, professor of biological sciences at Stanford University, and faculty director of Stanford's Jasper Ridge Biological Preserve. For most of the last two decades Dr. Field has fostered the emergence of global ecology. His research emphasizes ecological contributions across the range of Earthscience disciplines. Dr. Field and his colleagues have developed diverse approaches to quantifying large-scale ecosystem processes, using satellites, atmospheric data, models, and census data. They have explored local and global patterns of climate change impacts, vegetation-climate feedbacks, carbon cycle dynamics, primary production, forest management, and fire. For more than a decade, Dr. Field has led major experiments on grassland responses to global change, experiments that integrate approaches from molecular biology to remote sensing at the ecosystem-scale. His activities in building the culture of global ecology include service on many national and international committees, including committees of the National Research Council, the International Geosphere-Biosphere Programme, and the Earth System Science Partnership. Dr. Field was a coordinating lead author for the fourth assessment report of the Intergovernmental Panel on Climate Change. He is a fellow of the ESA Aldo Leopold Leadership Program and a member of the National Academy of Sciences. He has served on the editorial boards of Ecology, Ecological Applications, Ecosystems, Global Change Biology, and PNAS. Dr. Field received his Ph.D. from Stanford in 1981 and has been at the Carnegie Institution since 1984. His recent priorities include high performance "green" laboratories, integrity in the use of science by governments, local efforts to reduce carbon emissions, ecological impacts of biofuels, and the future of scientific publishing.

Donald F. Boesch
University of Maryland Center for Environmental Science

Donald F. Boesch is a professor of marine science and president of the University of Maryland Center for Environmental Science. From June 2002 through October 2003 Dr. Boesch also served as interim vice chancellor for academic affairs of the University System of Maryland. In 1980 he became the first executive director of the Louisiana Universities Marine Consortium, where he was also a professor of marine science at Louisiana State University. He assumed his present position in Maryland in 1990. Dr. Boesch is a biological oceanographer who has conducted research in coastal and continental-shelf environments along the Atlantic Coast and in the Gulf of Mexico, eastern Australia, and the East China Sea. He has published two books and more than 85 papers on marine benthos, estuaries, wetlands, continental shelves, oil pollution, nutrient overenrichment, environmental assessment and monitoring, and science policy. His research presently focuses on the use of science in ecosystem management. Dr. Boesch is active in extending knowledge to environmental and resource management at regional, national and international levels. He has served as science adviser to many state and federal agencies and regional, national, and international programs. He has chaired numerous committees and scientific assessment teams that have produced reports on a wide variety of coastal environmental issues. A native of New Orleans, Don Boesch received his B.S. from Tulane University and Ph.D. from the College of William and Mary. He was a Fulbright postdoctoral fellow at the University of Queensland and subsequently served on the faculty of the Virginia Institute of Marine Science.

F. Stuart "Terry" Chapin III
University of Alaska, Fairbanks

F. Stuart "Terry" Chapin III is a professor of ecology in the Institute of Arctic Biology at the University of Alaska, Fairbanks. Over the past 15 years Dr. Chapin has participated in several international commissions focused on the environment and climate, including the Millennium Ecosystem Assessment and Intergovernmental Panel on Climate Change. He was the first Alaskan elected to the National Academy of Sciences. Among the honors he has received are the Kempe Award for Distinguished Ecologist in 1996 and the Usabelli Award for the top researcher in all fields from the University of Alaska in 2000. Dr. Chapin received his Ph.D. in biological sciences from Stanford University in 1973.

Peter H. Gleick
Pacific Institute

Peter H. Gleick is the president of the Pacific Institute in Oakland, California, which he cofounded in 1987 and where he works on issues related to the environment, economic development, and international security, with a focus on global freshwater challenges. Among the issues he has addressed are conflicts over water resources, the impacts of climate change on water resources, the human right to water, and the problems of the billions of people without safe, affordable, and reliable water and sanitation. Dr. Gleick received a B.S. from Yale University and an M.S. and Ph.D. in energy and resources from the University of California, Berkeley. Dr. Gleick is the author of the biennial series on the state of the world's water, called The World's Water, published by Island Press, Washington, D.C., regularly provides testimony to the U.S. Congress and state legislatures, and has published many scientific articles. He served as lead author of the water sector report of the U.S. National Assessment of the impacts of climatic change and variability and as a reviewer for several chapters in the IPCC reports. In 2003 he was awarded a MacArthur Fellowship for his work on water resources and in 2006 he was elected fellow of the American Association for the Advancement of Science and to Membership in the National Academy of Sciences.

Anthony C. Janetos
University of Maryland

Anthony C. Janetos is director of the Joint Global Change Research Institute at the University of Maryland. Dr. Janetos previously served as vice president of the H. John Heinz III Center for Science, Economics and the Environment in Washington, D.C., where he directed the center's Global Change Program. He has written and spoken widely to policy, business, and scientific audiences on the need for scientific input and scientific assessment in the policy-making process and about the need to understand the scientific, environmental, economic, and policy linkages among the major global environmental issues. Dr. Janetos has served on several national and international study teams, including working as a cochair of the U.S. National Assessment of the Potential Consequences of Climate Variability and Change. He also was an author of the Intergovernmental Panel on Climate Change's Special Report on Land-Use Change and Forestry, the Global Biodiversity Assessment, and a coordinating lead author in the recently published Millennium Ecosystem Assessment. He currently serves as a member of the National Research Council's Committee on Earth Science and Applications from Space. Dr. Janetos graduated magna cum laude from Harvard College with a bachelor's degree in biology and earned a master's degree and a Ph.D. in biology from Princeton University.

Jane Lubchenco
Oregon State University

Jane Lubchenco is the Wayne and Gladys Valley Professor of Marine Biology and Distinguished Professor of Zoology at Oregon State University. Dr. Lubchenco is an environmental scientist and marine ecologist who is actively engaged in teaching, research, synthesis, and communication of scientific knowledge. Her expertise includes interactions between humans and the environment: biodiversity, climate change, sustainability science, ecosystem services, marine reserves, coastal marine ecosystems, the state of the oceans, and the state of the planet. She leads an interdisciplinary team of scientists who study the marine ecosystem off the west coast of the United States. This PISCO team is learning how the ecosystem works, how it is changing and how humans can modify their actions to ensure continued benefit from ocean ecosystems. She is a former president of the International Council for Science and a former president of the American Association for the Advancement of Science and the Ecological Society of America. She was a presidential appointee to two terms on the National Science Board which advises the President and Congress and oversees the National Science Foundation. She often testifies before Congress, addresses the United Nations, and provides scientific advice to the White House, federal and international agencies, nongovernmental organizations, religious leaders, and leaders of business and industry. In 1996, she was elected to the National Academy of Sciences, American Academy of Arts and Sciences, American Philosophical Society, Royal Society, and the Academy of Sciences for the Developing World. Dr. Lubchenco has received numerous awards, including a MacArthur Fellowship, a Pew Fellowship, and the 2005 American Association for the Advancement of Science's Award for Public Understanding of Science and Technology. She graduated from Colorado College, received her Ph.D. from Harvard University in marine ecology, taught at Harvard for two years, and has been on the faculty at Oregon State University since 1978.

Jonathan T. Overpeck
University of Arizona

Jonathan T. Overpeck is a professor of geosciences and director at the Institute for the Study of Planet Earth at the University of Arizona. His specialty areas are climate dynamics, including paleoclimatology, climate and ecosystem interaction, and climate assessment and decision support. Dr. Overpeck has written over 100 publications on climate and ecosystem variability, and was founding cochair of both international and U.S. CLIVAR-PAGES working groups. He is the chair of the National Science Foundation Arctic System Science Committee, and a member of the NOAA Climate Working Group. Dr. Overpeck has been awarded the U.S. Department of Commerce Gold Medal, as well as the Walter Orr Roberts Award of the American Meteorological Society for his interdisciplinary research. He is also a coordinating lead author for the ongoing Intergovernmental Panel on Climate Change Fourth Assessment. Most recently Dr. Overpeck was awarded a Guggenheim Fellowship to investigate paleoenvironmental perspectives on the future. He has a Ph.D. from Brown University.

Camille Parmesan
University of Texas

Camille Parmesan is associate professor of integrative biology at the University of Texas. Dr. Parmesan's early research focused on multiple aspects of population biology, including the ecology, evolution and behaviors of insect-plant interactions. For the past several years the focus of her work has been on current impacts of climate change on wildlife in the 20th century. Her work on butterfly range shifts has been highlighted in many scientific and popular press reports, such as in Science, Science News, New York Times, London Times, National Public Radio, and the recent BBC film series State of the Planet with David Attenborough. The intensification of global warming as an international issue led her into the interface of policy and science. Dr. Parmesan has given seminars for the White House, government agencies, and NGOs. As a lead author she was involved in multiple aspects of the Third Assessment Report of the Intergovernmental Panel on Climate Change. She received her Ph.D. at the University of Texas in 1995.

Terry L. Root
Stanford University

Terry L. Root is a senior fellow in the Woods Institute for the Environment, and professor by courtesy in the Biology Department at Stanford University. Dr. Root's research focuses on how wild animals and plants have responded over the last century when the global average temperature increased ~0.8°C and what the future ecological consequences will be for wild species as the globe continues to warm rapidly. She and coworkers have also used studies of species to show that humans are indeed causing a large portion of the increase in local and regional temperatures. As the planet continues to warm at an escalating rate more and more species will be at risk of extinction. Dr. Root's current work is focusing on how to avert as many extinctions as possible. She is working to help communicate information about climate change to decision makers and the general public. She was a lead author in both the third and fourth assessment reports of the Intergovernmental Panel on Climate Change Working Group 2. In 2007 the IPCC shared the Nobel Peace Prize with Vice President Gore. In 1990 she received the Presidential Young Investigator Award from the National Science Foundation and in 1992 was selected as a Pew scholar in Conservation and the Environment and Aldo Leopold leadership fellow in 1999. She received her bachelors degree in mathematics and statistics at the University of New Mexico, her masters degree in biology at the University of Colorado in 1982, and her Ph.D. in biology from Princeton University in 1987. She has served on the National Research Council Committee on Environmental Indicators and Pierce's Disease in Vineyards of California

Steven W. Running
University of Montana

Steven W. Running is trained as a terrestrial ecologist, receiving B.S. (1972) and M.S. (1973) degrees from Oregon State University and a Ph.D. (1979) in forest ecology from Colorado State University. Since 1979 he has been with the University of Montana, Missoula, where he is a University Regents Professor of Ecology. His primary research interest is the development of global and regional ecosystem biogeochemical models by integration of remote sensing with climatology and terrestrial ecology. He is a team member for the NASA Earth Observing System, Moderate Resolution Imaging Spectroradiometer, and he is responsible for the EOS global terrestrial net primary production and evaporative index datasets. He has published over 240 scientific articles. Dr. Running has recently served on the standing Committee for Earth Studies of the National Research Council and on the federal Interagency Carbon Cycle Science Committee. He is a cochair of the National Center for Atmospheric Research Community Climate System Model Land Working Group, a member of the International Geosphere-Biosphere Program Executive Committee, and the World Climate Research Program, Global Terrestrial Observing System. Dr. Running shared the Nobel Peace Prize in 2007 as a chapter lead author for the Fourth Assessment of the Intergovernmental Panel on Climate Change. He is an elected Fellow of the American Geophysical Union and is designated a highly cited researcher by the Institute for Scientific Information.

Stephen H. Schneider
Stanford University

Stephen H. Schneider is a professor in the Department of Biological Sciences, a senior fellow at the Woods Institute for the Environment, and professor by courtesy in the Department of Civil and Environmental Engineering at Stanford. In 1975 he founded the interdisciplinary journal, Climatic Change, and continues to serve as its editor. Dr. Schneider's current global change research interests include climatic change, global warming, food and climate and other environmental science public policy issues, ecological and economic implications of climatic change, integrated assessment of global change, climatic modeling of paleoclimates and of human impacts on climate (carbon dioxide greenhouse effect and environmental consequences of nuclear war, for example). He is also interested in advancing public understanding of science and in improving formal environmental education in primary and secondary schools. He was honored in 1992 with a MacArthur Fellowship for his ability to integrate and interpret the results of global climate research through public lectures, seminars, classroom teaching, environmental assessment committees, media appearances, congressional testimony, and research collaboration with colleagues. He has served as a consultant to federal agencies and White House staff in the Nixon, Carter, Reagan, Bush Sr., Clinton, and Bush Jr. administrations. He received, in 1991, the American Association for the Advancement of Science Westinghouse Award for Public Understanding of Science and Technology for furthering public understanding of environmental science and its implications for public policy. He was elected to membership in the National Academy of Sciences in 2002. Dr. Schneider received his Ph.D. in mechanical engineering and plasma physics from Columbia University in 1971.